CARBONATE STONE

CARBONATE STONE
CHEMICAL BEHAVIOR, DURABILITY, AND CONSERVATION

K. Lal Gauri
Jayanta K. Bandyopadhyay
University of Louisville
Louisville, KY

A WILEY-INTERSCIENCE PUBLICATION
JOHN WILEY & SONS, INC.
New York · Chichester · Weinheim · Brisbane · Singapore · Toronto

For ordering and customer service, call 1-800-CALL-WILEY.

Library of Congress Cataloging-in-Publication Data

Gauri, K. Lal.
    Chemical analysis, decay, and conservation of stone monuments / K. Lal Gauri, Jayanta K. Bandyopadhyay.
        p. cm
    "A Wiley-Interscience publication."
    ISBN 0-471-17977-9 (alk. paper)
    1. Building stones—Deterioration.   2. Monuments—Conservation and restoration.   3. Building stones—Analysis.   I. Bandyopadhyay, Jayanta K. II. Title
TA427.G38 1999
691'.2—dc21   99-25453
                    CIP

Printed in the United States of America.

10 9 8 7 6 5 4 3 2 1

◼◼◼ CONTENTS

FOREWORD

That which man has for centuries created out of the most durable material at hand, the rocks around him, is, of course, carved in stone, and provides us with the best record of the very distant past. It would seem to be as long lasting and inevitable as time itself. These are not just ancient relics but part of our present culture, the natural ambience of history. As Joseph Polzer wrote in 1982 in his introduction to the papers of the Fourth International Congress on the Deterioration and Preservation of Stone Objects, a meeting that was organized in Louisville by this book's author, K. Lal Gauri,

> Even when outstanding historic structures are isolated as symbols or records of distant cultures in museums in the open, such as the Parthenon on its acropolis, the Taj Mahal in its park, or the pyramids of Egypt rising from desert sands, they recreate the past for us with an immediacy that no mental image can match.

Yet eventually stone, all stone, deteriorates and disappears. This is well known. The problems have rapidly become more alarming as the world has become industrialized and polluted. Programs are being developed to find ways of stemming the loss, which is an odd but important aspect of the preservation of the information age.

Greek and Roman writers told of treatments to forestall stone deterioration centuries ago. So-called scientific approaches began to appear in the nineteenth century. Concern for the preservation of archaeological treasures excavated in Egypt, Pompeii, and other sites in the Middle East brought more effort to the task early in the twentieth century. Both the analytical tools needed to develop an understanding of the processes of deterioration and the availability of appropriate chemicals and treatment methods were lacking. Saving deteriorating stone was primarily a matter of applying whatever wax, paint, or other coating was available and might work. After about 1960, polymeric materials with more suitable properties became available commercially, and these were applied as the need dictated. They were effective but rarely provided the long-term stabilization that is needed for objects standing outdoors. More recently, an urgency has developed for us to use our scientific tools to better understand the nature of the stone, the causes of failure, and the mechanisms of deterioration before defining a treatment. Clifford Price in 1996 eloquently stated these requirements in *Stone Conservation: An Overview of Current Research*:

> If we are to do anything to reduce or prevent this loss of our heritage we must first be able to characterize the many stones involved. We need to be able to describe the decay, and to measure the extent and severity of decay. We need to understand the causes and mechanisms of decay. Only then can we hope to understand the behavior of any particular stone in a given environment.

This book fills the prescription urged by Price to cure the diseases of stone, which is to provide an accurate diagnosis before treatment. K. Lal Gauri has chosen carbonate rocks as the stone to be addressed. He has spent a career studying these materials. They are the elements of widespread and major architectural, statuary, and decorative usage, yet as derivatives of carbonate minerals, they are uniquely susceptible to rapid degradation by the acid gases found in modern environments. Price requires that we first be able to characterize the stone. This is covered in chapters on the origin, occurrence, properties, and classification of carbonate rocks, and the impact of noncarbonate minerals when they are present. Decay is described in sections on the weathering of calcareous stone both in natural (without acidic industrial gases) and polluting atmospheres, and in a chapter on biodeterioration. The reader will learn how to measure the extent and the severity of decay in chapters on porosimetry and the determination of decay rates. These combine to provide an understanding of the causes and mechanisms of decay which is illustrated in chapters on the deterioration and durability of the Sphinx. Finally, an example of how this provides a basis for treating a building under deterioration and preserving it as an active cultural center is offered in a chapter describing Gauri's work on the preservation of the California Building in San Diego. The well-attended usage of the California Building as an anthropological museum is a fitting tribute to Lal Gauri's productive career.

CHARLES M. SELWITZ

The Getty Conservation Institute
Los Angeles

This book is designed for students of conservation technology, which has recently become an independent discipline at many universities and technical schools around the world. It can be useful also to architects, civil engineers, and urban planners—in developing geographic sites and structural designs—giving the description, durability, and sources of carbonate rocks. In addition, chemical engineers will find it useful for its treatment of modeling gas–solid and aqueous carbonate reactions and geologists for quantitative treatment of modern weathering.

Carbonate rocks include limestone, dolostone, and marble and are composed largely of the minerals calcite ($CaCO_3$) and dolomite [$CaMg(CO_3)_2$]. Limestone is the parent rock from which dolostone is formed by replacing calcium with magnesium in a diagenetic environment. Marble is formed by recrystallizing calcite or dolomite in metamorphic terrain. In Chapter 1 we discuss the origin, properties, identification, and classification of these rocks, the knowledge of which is essential for understanding their behavior when exposed outdoors as buildings, monuments, and sites.

Limestone is a sedimentary rock deposited as a soft sediment on shallow floors of tropical seas. During sedimentation some noncarbonate minerals are washed into carbonate basins and become part of the limestone. These minerals contribute significantly to the weathering of the carbonate rocks. We have devoted Chapter 2 to the origin and identification of these noncarbonate minerals. X-ray diffraction is the commonly applied technique for their identification. Hence, we have introduced in this chapter the subject of X-ray crystallography as applied to the identification of these minerals as well as other minerals discussed elsewhere in this book.

Epeirogenic movements, which cause an uplift of geological regions, heave the sediments above the sea floor. As a result, the sediments become hardened into rocks and acquire fractures or joints. The joints are an expression of what is called *structural deformation*. In Chapter 3 we describe methods of representing joints in stereographic projections whereby spatial preferred orientations, if present, can be discerned; in Chapter 9 we discuss the influence of joints on weathering of the Sphinx limestone. In Chapter 3 we also describe experiments that simulate structural deformation by crystallizing sodium sulfate in rock pores and measuring changes in strength properties, such as compressional strength and modulus of elasticity, by routine engineering techniques. We then develop mathematical correlation by means of which these properties can be determined without subjecting the rock to testing by engineering techniques.

The remainder of this book deals with chemical and physical weathering. We show how to analyze the different aspects of weathering and present mathematical models that can be used in determining the effects of weathering under different conditions.

The relationships thus developed facilitate predictions of chemical changes where reaction conditions are known and enable us to reconstruct past conditions from the reaction product.

Our investigations concerning chemical weathering confirm the common belief that the chemical decay of carbonate monuments is essentially a phenomenon of the twentieth century and is most prevalent where industrialization is highly advanced. Among the pollutant gases in industrial environments produced by combustion of fossil fuels, SO_2 and NO_x are the most potent. These gases react with stone that is protected from rain, producing crust which eventually exfoliates; surfaces that are washed by rain erode from the corrosive action of the acid rain. In environments where these gases are absent, carbon dioxide is, as it was in the past, the only chemically active gas, but CO_2 does not react with calcite in the gaseous phase so rain-sheltered surfaces are always protected. However, due to its low reactivity, even when dissolved in water, exposed surfaces had been, as now, weakly affected.

Thus, from the point of view of environmental conditions, world areas can be classified as polluted or nonpolluted; the latter are also said to have a natural environment. Chapters 4 and 5 deal, respectively, with weathering in natural and polluted environments. Chapter 6 deals with the mathematical modeling of chemical reactions in polluted environments. To cover the subject of weathering completely, we describe biodeterioration in Chapter 7.

Mechanisms of physical weathering are the same in both polluted and natural environments. They are related to stresses generated within the rock by water intrusion, expansion of clay minerals, and crystallization of water-soluble salts. The magnitude of stress produced, however, is controlled by pore-size distributions. These mechanisms are described in Chapter 4.

Durability of rock is understood to relate to its resistance to physical weathering, which is also controlled by porosity. Chapter 8 gives the measurement of pores by mercury porosimetry and describes various techniques, such as pore-potential and fractal dimension, whereby pore-size distributions can be used to construct durability factors. In Chapter 9 we discuss durability factors for the limestone strata of the Great Sphinx.

Ancient monuments are a cultural patrimony that must be conserved. These, and even modern structures, have often weathered to such an extent that conservation and restoration is needed. Chapter 10 discusses the diagnoses of the maladies of weathered stone and identifies some selection and performance specifications for treatments. We include in this chapter the case histories of treatments we designed for the California Building and the Sphinx.

Weathering has a dimension of time, and some scientists have attempted to determine the age of anthropological objects from the mode and the degree of their weathering. In Chapter 11 we discuss the controversial question of the age of the Sphinx.

Numerical data analysis is an essential constituent of any scientific study. In Appendix A we present the mathematical development of artificial neural networks (ANN), a very powerful tool for recognizing patterns in data, which we used in Chapter 6 and 9 for modeling reaction rates and durability factors. Similarly, many weathering problems relate to the condensation of moisture on stone surfaces. Appendix B gives relationships between temperature, relative humidity, and dew point.

A special feature of this book is that it treats all aspects of weathering and conservation quantitatively based on experiments that we conducted in laboratory and natural conditions. The subject of these experiments were natural rocks which are not uniform in composition and fabric. Consequently, the data often showed large variations from sample to sample even when the conditions of experiments were held constant. The mathematical relations developed from these data, therefore, when used to predict an unknown can produce a result that may be quite different from the average. An ftp site can be accessed through the Internet (see Appendix C) to solve problems when appropriate data are available.

During the preparation of this book we have received indispensable help from our families and friends. Special thanks are given to our wives, Kamla and Sailaja, without whose understanding, support, and patience this book could not have been written. Thom Lierman took many micrographs showing carbonate petrology as well as scanner images of illustrations. John Sinai devoted many hours to reviewing the manuscript. Our highest gratitude goes to the National Science Foundation whose financial support over the years has made the research possible on which this book is based.

K. LAL GAURI

JAYANTA K. BANDYOPADHYAY

University of Louisville

CARBONATE STONE

CHAPTER 1

Origin, Occurrence, Properties, and Classification of Carbonate Rocks

1.1 INTRODUCTION

Carbonate rocks appear in three basic forms: limestone, marble, and dolostone. By definition, these rocks contain more than 50% carbonate minerals. They differ from one another in mineral composition and texture. The most common carbonate rock is limestone, composed of calcium carbonate ($CaCO_3$) in the form of the mineral calcite. Another abundant carbonate rock is dolostone. The common mineral of dolostone is dolomite, $CaMg(CO_3)_2$. Marble, perhaps the most commonly used rock in statuary, is made of calcite or dolomite. Most famous marbles, such as the Carrara marble of Italy, the Penetelic marble of Greece, and the Makrana marble of India, however, are all made of calcite. Limestone is the parent of all carbonate rocks; marble and dolostone are formed by alteration of limestone.

Carbonate rocks have been and continue to be the major source of architectural and statuary stone. The pyramids, the Sphinx, and most other Egyptian monuments were built from limestone; the Taj Mahal (Fig. 1.1) in India and most Greek and Roman architecture, including the many temples at Acropolis, were built of marble. Famous pieces of statuary in dolostone are also well known, but the primary importance of dolostone has been in common construction.

The major objective of this chapter is to characterize all carbonate rocks, because different rock types weather differently and at different rates when exposed to the atmosphere. Thus the knowledge of the rock type can give an indication of its behavior when it is used in the construction of masonry structures and monuments. Alternatively, the identification of the original rock from existing structures and the knowledge of the prevailing atmosphere allow understanding of mechanisms involved in the deterioration processes. This understanding can provide methods for treating weathered materials to achieve soundness.

Composition and texture are the main properties that distinguish one rock type from another. These properties, in particular, the depositional texture of limestone—by far the most common carbonate rock—result from the mode of origin of the carbonate sediment and the changes the sediments undergo when they indurate to hard rock. We will first describe the identification of carbonate minerals compared to noncarbonate minerals and then the identification of individual carbonate minerals. Then we will classify carbonate sediments based on their primary depositional texture

1

FIGURE 1.1 Taj Mahal, Agra, India. Taj Mahal is made of marble, a carbonate rock used in many famous architectural works all over the world.

and limestone types on the basis of preserved depositional texture. We will follow this with the study of dolostone and marble, which are formed by the replacement and recrystallization of limestone.

1.2 IDENTIFICATION OF CARBONATE MINERALS

Limestone and marble can be easily distinguished from noncarbonate rocks by dropping dilute (5% or less, by volume) hydrochloric acid on the sample. Calcite dissolves in cold HCl by effervescence, or bubbling, caused by the release of carbon dioxide:

$$CaCO_3 + 2HCl \rightarrow Ca^{2+} + 2Cl^- + CO_2 + H_2O \qquad (1.1)$$

Visible effervescence can also be produced by the reaction of HCl with dolomite. However, the acid must be more concentrated and warm.

In the preceding test the acid etches the stone components preferentially. As a result, various textural components of limestone can be easily recognized. While a detailed discussion of this matter is presented in the following, suffice it to say here that dolomite can be easily distinguished from limestone or marble by the

TABLE 1.1   Mohs' Hardness Scale of Minerals[a]

1. Talc	6. Feldspar *File*
2. Gypsum *Fingernail*	7. Quartz
3. Calcite *Copper coin*	8. Topaz
4. Fluorite	9. Corundum
5. Apatite *Knife blade, glass*	10. Diamond

[a]The Mohs scale is an empirical scale because it is not known by which scale (linear, fractional, exponential) one group of minerals is related to the other. Also, the underlying cause of the hardness (atomic properties, crystal structure etc.) is not quantitatively correlated with hardness.

rhombohedral etch patterns that show the typical outline of the dolomite grains. Also, weathered dolomite shows brown coloration due to the oxidation of ferrous iron, which is often present in trace quantities in dolostone. The specific gravities of calcite and dolomite are also different: 2.72 for calcite and 2.85 for dolomite. Staining is another common technique to distinguish these two minerals: Calcite grains stain pink with alizarin red S, whereas dolomite grains do not stain. Furthermore, calcite has a hardness of 3 on the Mohs scale[*] (Table 1.1). Limestone and marble surfaces therefore can be scratched with a copper coin. The hardness of dolomite is 3.5 to 4; a copper coin will not scratch it but a pocketknife or glass will.

If an X-ray diffractometer is available, diffraction traces are the most convenient means to distinguish calcite from dolomite (Fig. 1.2). A detailed account of X-ray diffraction is given in Sec. 2.2.3.2.

Finally, limestone and dolostone are porous rocks, whereas marble is nearly nonporous. Methods to determine porosity are given in Secs. 5.2.3 and 8.3.

1.3   CARBONATE SEDIMENTS

Carbonate sediments are precursors to all carbonate rocks. Limestone forms by their consolidation, dolostone by their replacement, and marble by their metamorphism. First, we will describe carbonate sediments followed by the study of limestone, dolostone, and marble.

1.3.1   Origin

Carbonate sediments are deposited from warm shallow waters on the shelves of tropical seas. Marine water is supersaturated with respect to calcium (Ca^{2+}) and carbonate (CO_3^{2-}) ions, derived mostly by the weathering of limestone on land and

*Mohs' hardness scale is based on the relative "scratchability," that is, a harder mineral is able to scratch a groove on a smooth fresh surface of a softer mineral. For example, quartz can always scratch calcite. The Mohs scale can be augmented by using common objects to test for hardness. A fingernail usually has a hardness between 3 and 4 and a knife blade or a steel nail file has a hardness slightly greater than 5.

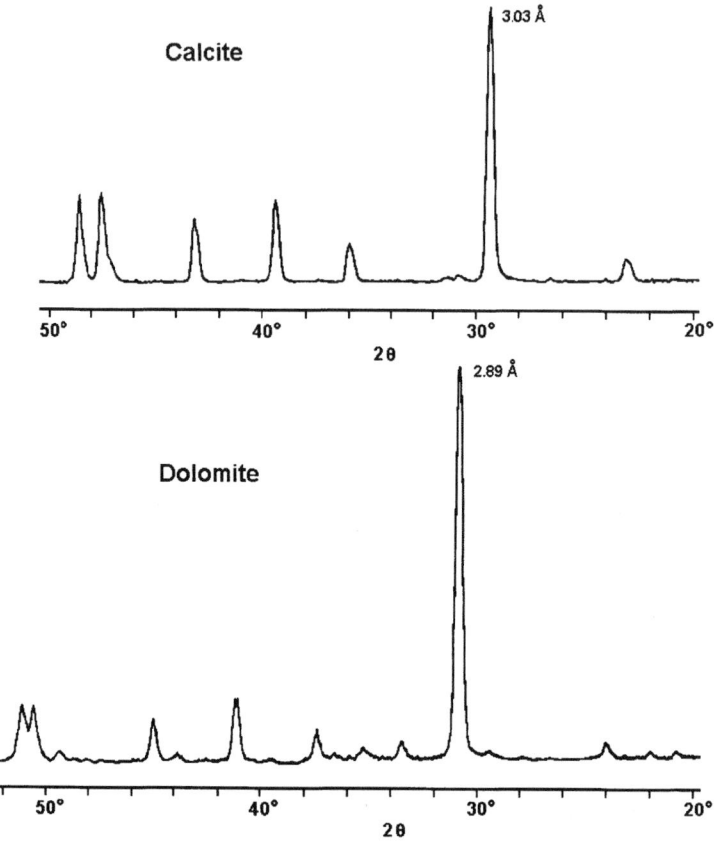

FIGURE 1.2 X-ray diffraction (XRD) traces of calcite (top) and dolomite (bottom), the most common carbonate minerals. A detailed treatment of XRD and its application in identification of minerals is given in Chapter 2.

subsequent transportation to the sea or world oceans by rivers. Seawater is in fact river water that has become concentrated in salts due to evaporation over geologic time.

The solubility of calcium and carbonate ions in water is a function of the amount of dissolved carbon dioxide. A slight reduction of carbon dioxide in water can trigger the precipitation of calcium carbonate. Normally occurring biochemical and chemical processes may bring about this reduction. Thus, the formation of carbonate sediment is a rather common phenomenon as shown by the abundance of limestone throughout geologic time and space. In the United States, for example, limestone covers a vast tract of land from the Appalachian to the Rocky Mountains, spanning geologic time from the early Paleozoic era to the Recent epoch, more than 500 million years (Fig. 1.3).

In modern sediments, aragonite and the high- and low-magnesian calcite are the primary minerals that form in order of abundance. The chemical formula for these

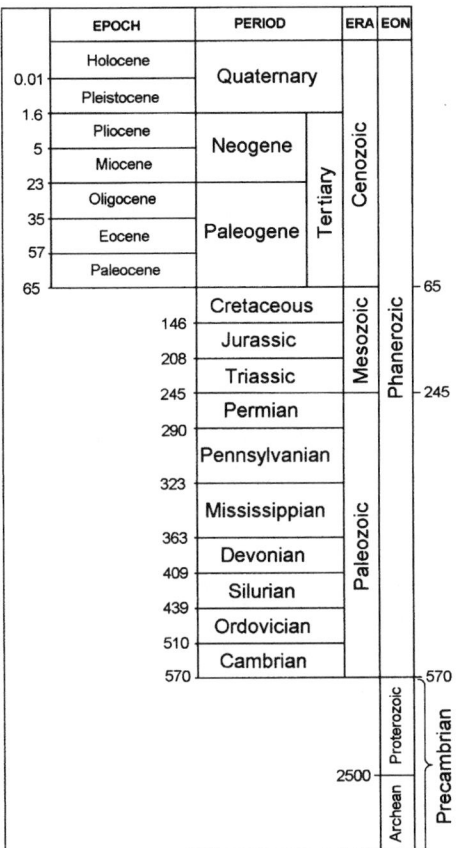

FIGURE 1.3 Geologic time scale. The geologic time scale of the earth is based on the paleontologic and stratigraphic record. The ages (given in million years) are, however, estimates based on the record of radioactive elements trapped in igneous rocks. Carbonate rocks span a long geologic time. The Makrana marble of the Taj Mahal (Fig. 1.1) is of Proterozoic age. The limestone and dolomite types mentioned in this chapter range from Silurian period to Eocene epoch.

minerals is expressed as $CaCO_3$. Aragonite is a polymorph of calcite, that is, it has the same composition as calcite but a different crystal structure. High- and low-magnesian calcite is distinguished by a magnesium carbonate ($MgCO_3$) content of more and less than 4%, respectively. Carbonate sediment changes into limestone rock by diagenesis (discussed later). However, all limestone rocks contain pure calcite because aragonite and magnesian calcite are unstable in the environment under which the carbonate sediment changes to rock.

Even though chemical processes are at work, carbonate sedimentation is almost entirely due to biochemical activity. Some examples by which this activity leads to the formation of carbonate sediment include the following.

- Many marine organisms have skeletons constructed of calcium carbonate. When these organisms die they settle to the sea floor, and the skeletons are worked by waves and often broken down to fragments of variable sizes. The white sand beaches of the tropical seas are a manifestation of this process. Such an accumulation of shell fragments is only one of several types of carbonate sediments. Another type is formed when waves are absent in a sheltered seashore and skeletal organisms attached to the sea floor become buried. Furthermore, many blue-green algae grow as mats on the floor of intertidal and subtidal zones. The algae cells contain aragonite needles that become separate with decomposition of the soft tissue. These aragonite needles are the main constituents of carbonate mud found ubiquitously on the modern shallow tropical sea floor.

- Many bacteria are able to precipitate calcium carbonate from seawater both in and outside the cell wall, which then collects in a large mass to make sediment. Folk (1997) considers that most carbonate sedimentation may have been a result of work by small bacteria (0.05 μm to 0.2 μm) termed by him as *nannobacteria*.

- Many organisms with variable architectural functions construct reef structures, termed *bioherms*, on the sea floor. Thus, the reef in itself is an accumulation of carbonate sediment.

- Carbonate sediment may also form indirectly as a result of biologic activity. For example, photosynthesis by algae reduces the carbon dioxide content of seawater. Thereby, the solubility of calcium and carbonate ions is reduced. This results in the precipitation of calcium carbonate. This is a chemical precipitate due to a biochemical process.

Nevertheless, some purely chemical precipitation of calcium carbonate also occurs, induced by a higher water temperature at which the solubility of carbon dioxide in water is reduced. Since the solubility of calcium carbonate is directly related to the carbon dioxide concentration, its release causes the calcium carbonate to precipitate.

1.3.2 Texture of Carbonate Sediments

In both modern carbonate sediments and limestone rocks, texture refers to the geometrical features of component particles including size, shape, and arrangement. The depositional textures of the sediments are often preserved in the limestone. Thus, the study of modern sediments aids in the understanding of ancient limestone textures. The textural components of carbonate sediments are fine-grained mud, coarse-grained particles, and crystalline cement (Fig. 1.4).

Fine-Grained Mud. Fine-grained mud is also termed *micrite*. As a modern sediment it is a microcrystalline ooze of clay size [≤ 4 μm (1 μm = 10^{-4} cm)] containing aragonite needles that are embedded in the soft tissue of certain algae. With disintegration of the soft tissue, these needles are recrystallized into tiny particles, which make up the main mass of the sediment. When some

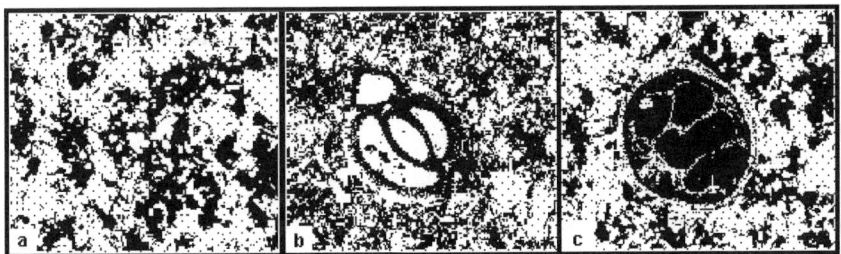

FIGURE 1.4   Carbonate sediment textures. In carbonate sediments texture refers to size and nature of particle. Particles less than 4 μm are called microcrystalline (a); those over 10 μm are allochems, shown by a shell (b); and sparry calcite, shown by crystalline fringes around a shell (c). Sediments composed largely of microcrystalline material are called biomicrites and those made largely of large particles are called allochemical sediments. Sparry calcite, rarely present in sediments, is considered to form by chemical precipitation of calcite, whereas other textural components are largely of biochemical origin.

coarse-grained particles are present, micrite is said to make the matrix of the sediment.

Coarse-Grained Particles. Coarse-grained particles are also called *allochemical particles* or simply *allochems*. They are discrete carbonate particles of various kinds, generally larger than 10 μm. The sources of coarse-grained particles are shell fragments, ooids, pellets, and interclasts, described later.

Shell fragments. As stated previously, most marine organisms have their skeletons made of calcium carbonate. Wave action breaks up these shells and sorts the fragments according to particle size. As well-sorted sediment, carbonate sand contains particles of size 0.6 mm to 2 mm. When not well sorted, particles of carbonate sand may be present in otherwise micritic sediment.

Ooids. Ooids are elliptical or spherical particles with a concentric or radial multilayered fabric. The nucleus of the ooids is generally a minute shell fragment. In modern sediment, the concentric rims of ooids are made of aragonite needles, which may be secreted, in situ, by bacteria, obtained as a coating from the carbonate mud or formed as a pure chemical precipitate.

Pellets. Pellets are small spherical particles, 10 μm to 30 μm diameter, of microcrystalline carbonate without internal fabric. Often they are the fecal products of organisms that ingest the carbonate mud for the organic matter.

Intraclasts. Intraclasts are fragments of somewhat consolidated carbonate sediment that was reworked by the waves. Their identification is important from genetic point of view as it reveals that the sediment had formed in a high-energy environment, explained later.

Sparry calcite. Sparry calcite consists of chemically precipitated clear calcite. Sparry calcite is well developed in rocks but is rarely found in sediments.

Until the 1950s, carbonate mud was considered to be an inorganic precipitate. Earlier works used the term *orthochemical* for the micrite and sparry calcite and considered the shell fragments only as being biochemical.

1.3.3 Environment of Deposition

Environment of deposition refers to hydraulic conditions. Certain coastal shelves are washed by strong waves, while others are covered by perfectly calm water. The wave action, or the absence of it, controls the texture of the sediment.

A high-energy environment is said to exist with brisk water activity. This water may carry particles of varying sizes, but only those that are too large to be carried by the current will settle. The limit of the smallest particles settling in a high-energy environment is placed at 10 µm. Thus, allochemical sediment is deposited in a high-energy environment.

Particles of micrite, that is, particles of less than 4 µm diameter, will be winnowed away from the high-energy environment and deposited on the floor of a calm portion of the sea, such as the shelf of a lagoon, in which large particles could not have originally been present because of the lack of wave activity.

Thus, in the calm environment of a lagoon or on other sheltered portions of the seashore, micrite can deposit and remain on the sea floor. The presence of allochemical particles is possible, however, in such an environment because such particles may be washed in by a storm or organisms living on the sea floor may be buried. Consequently, when micrite is present, a spectrum of sediment concentration exists ranging at one end from a purely micritic sediment to the other end at which allochemical particles predominate, but with some micrite enclosed between the particles.

The third textural element, sparry calcite, forms independently of the hydraulic conditions as it crystallizes from seawater trapped in the pores of the sediment. It is nearly absent in micritic sediment, but allochemical sediment may contain some sparry calcite in the form of fibrous coatings on grain surfaces.

1.4 LIMESTONE

Soft carbonate sediments convert to hard limestone over time. The processes through which sediments are converted to hard rock are included in what is called diagenesis.

1.4.1 Diagenesis

Diagenetic processes are at work when sediment accumulates on the sea floor as well as when sediment is lifted above the sea floor. These processes include compaction, recrystallization, cementation, and replacement. Calcite may also be dissolved and then sometimes removed during diagenesis; however, we will consider this later in relation to porosity.

- *Compaction.* Compaction is the partial consolidation of sediment due to the pressure exerted by the overlying sediment. Micritic carbonate sediments experience maximum compaction because of the tendency of fine grains to become well packed.

- *Cementation.* Cementation is the process whereby inorganically precipitated primary calcite recrystallizes and bonds the grains. This recrystallized calcite is called sparry calcite. Sparry calcite is found in a predominantly allochemical limestone because the parent sediment is grain supported, that is, the mechanical structure is supported due to grains touching each other, and in order to convert to a rock the grains must be bonded with a cement. The mineral solution contained in the pores of allochem-rich sediment supply the calcite. Nonetheless, a minor quantity of microspar cement may also be found in micritic rocks, which are said to be mud supported.

- *Recrystallization.* Recrystallization of carbonate sediment refers to the conversion of the first-formed carbonate minerals aragonite and Mg-rich calcite to pure calcite. As indicated previously, the first two minerals mentioned are unstable in a nonmarine environment due to the ionic composition of the freshwater. Freshwater has a lower Mg:Ca ratio than that of seawater. Thus, when freshwater is able to move through the sediment, commonly when the sediment has been lifted above sea level, recrystallization occurs. Recrystallization also refers to the postdiagenitic secondary crystallization of calcite. Recrystallization mostly occurs in ancient rocks once deeply buried where great pressure dissolved calcium carbonate (pressure solution) in one place and deposited in pores elsewhere. This calcite, like the primary cement discussed previously, is also called sparry calcite.

- *Replacement.* Replacement is the same process as recrystallization; however, it implies the partial replacement of calcium in calcite, especially by magnesium, when the new mineral dolomite, $CaMg(CO_3)_2$, is formed.

1.4.2 Texture of Limestone

The textural components of limestone can be best recognized by light microscopy of etched surfaces, peels thereof, or thin sections. Etching is performed by dropping dilute (5% or less) hydrochloric acid on a polished surface for 1 or 2 minutes and rinsing the surface with water. Moistening an acetate film with a solvent, such as acetone, and pressing it against the etched surface make a replica of the etched surface. When dry, the film is peeled off and placed in a frame between two plates of glass to avoid curling. Thin sections are glass slides with slivers of mineral so thin that light can pass through them. They are made by cutting a chip from the rock, polishing a cut surface, and gluing the chip to the microscope slide. The chip is then ground to a thickness of nearly 30 μm.

The textural components have variable properties in transmitted and reflected light. Fine-grained mud is dull and opaque. Coarse grains are mostly fragments of large shells. They and the crystalline cement have vitreous luster in reflected light and are nearly transparent in transmitted light. Recognition of these components is essential in the study of carbonate rocks because the classification and description of these rocks are often based on texture.

1.4.3 Classification

In a series of publications, Folk (1962) and Dunham (1962), among others, recognized that most limestone consists of micrite, various types of allochems as detailed in the limestone types given in the following, and sparry calcite. Whereas micrite and allochems are the original constituents of sediment, sparry calcite, a constituent of allochemical rock, forms largely in the diagenetic process. With this background the following limestone types are recognized (Folk, 1962).

> *Type I Limestone.* This is microcrytalline limestone consisting entirely of ooze without the presence of allochems. An example is lithographic limestone (Fig. 1.5), which, because of its fine grain and ease to carve, was used in the ancient art of printing.

> *Type II Limestone.* Designated as microcrystalline allochemical limestone, it is further distinguished by the quantity of allochem present, varying from nearly 80% to almost 0%. Sparry calcite in such rocks is subordinate or lacking. The subtypes of this limestone are *sparse biomicrite* (Fig. 1.6), limestone with less than 10% allochems, and *packed biomicrite* (Fig. 1.7), limestone containing more than 10% allochems. These micrite limestone subtypes may be distinguished further by the nature of the allochem. For example, if ooids mainly make up the allochem, then the name given is *oomicrite*.

> *Type III Limestone.* Designated as sparry allochemical rock, it consists chiefly of allochemical grains cemented by sparry calcite cement. Depending on the type of the allochem present, the limestone is named *biosparite* (Fig. 1.8–1.11), *oosparite* (Fig. 1.12), *pellsparite* (Fig. 1.13), and *intrasparite* (Fig. 1.8), when shells, oolites, pellets, and intraclasts, respectively, are abundant.

> *Type IV Limestone.* A special type of limestone formed by the massive growth of organisms. These rocks make reef structures, also known as bioherms, and have an inherent rigid framework structure.

Figures 1.5 to 1.13 are micrographs of thin sections made under an optical microscope. Figures 1.14 and 1.15 show some of the rock types from which the thin sections were made. The classification of limestone is shown in Table 1.2.

Not included in this classification are the deep-sea carbonate ooze deposits consisting mainly of shells of foraminifera and other microscopic skeletons. *Chalk* is such a limestone. Because of its less well-compacted nature this limestone is not commonly used in masonry construction. Also excluded from this classification is *travertine*, which will be discussed toward the end of this chapter.

FIGURE 1.5 Lithographic limestone. This micrograph shows Solnhofen limestone, made entirely of microcrystalline calcite. Solnhofen limestone is well known for the Jurassic fossil *Archaeopteryx*, which was found in an excellent state of preservation because of the fine nature of the constituent particles and the fact that this limestone formed in calm water mostly devoid of oxygen. This and other lithographic limestones were used earlier in printing because of details that can be carved in a fine-grained material. Lithographic limestone is considered to have formed in a low-energy environment.

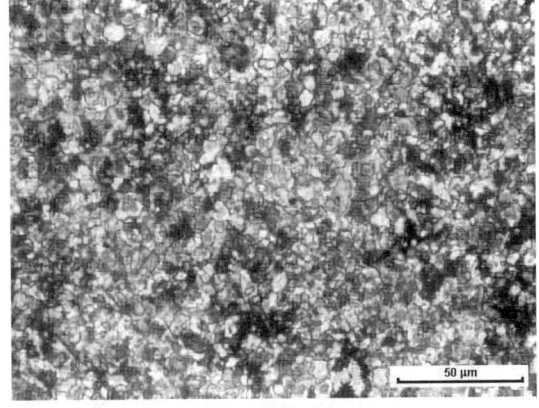

FIGURE 1.6 Sparse biomicrite or fossiliferous mudstone. This micrograph of a limestone sample from lower beds of Eocene age at the Sphinx shows a micritic matrix in which some shells are scattered. The large shell in the center of the micrograph is a gastropod shell, the chambers of which are filled with secondary sparry calcite. Scattered dark spots represent finely disseminated iron oxide.

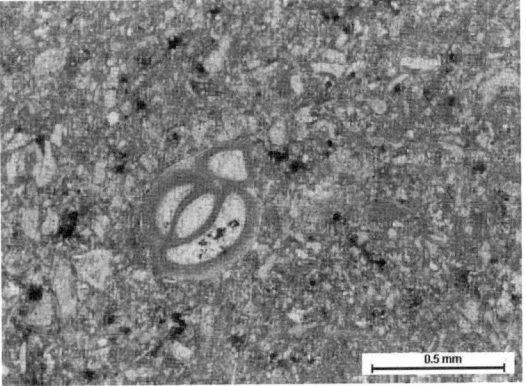

FIGURE 1.7 Packed biomicrite or packstone. This micrograph of a limestone sample from an upper bed of Eocene age at the Sphinx shows a large volume of allochem consisting mainly of skeletons of foraminifera.

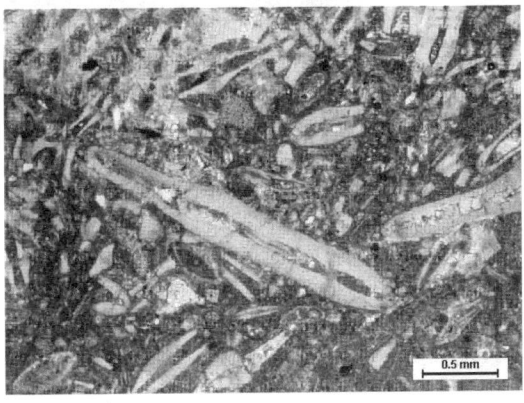

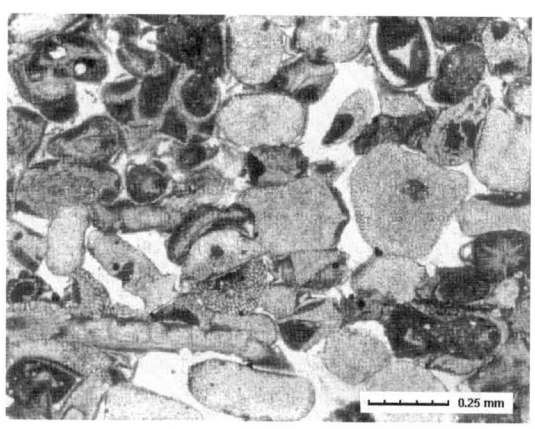

FIGURE 1.8 Biosparite or fossiliferous grainstone. This micrograph of a sample of Indiana limestone of Mississippian age shows that it is exclusively made of fossil shells cemented by sparry calcite. Darker areas in the intergranular space are the pores; the bright area represents sparry calcite. Sparites or grainstones are considered to have formed from high-energy carbonate sediment. Because of its high durability, Indiana limestone is the most widely used rock in architecture in the United States.

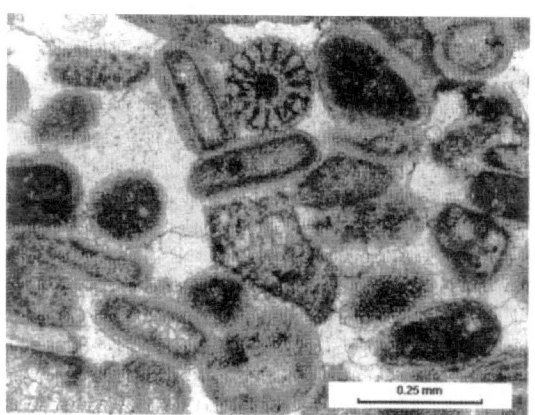

FIGURE 1.9 Biosparite. This micrograph is from a sample of the limestone Cordova Cream of Cretaceous age from Texas. Cordova Cream is similar to Indiana limestone (Fig. 1.8), but the grains are much smaller.

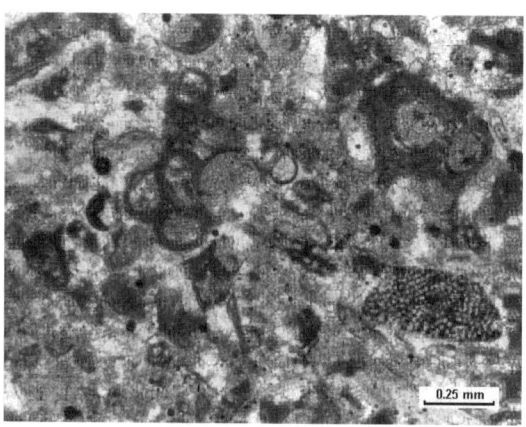

FIGURE 1.10 Poorly washed biosparite. This micrograph of a sample of the Leuders limestone of Permian age from Texas shows the presence of some micrite in an otherwise allochemical mass.

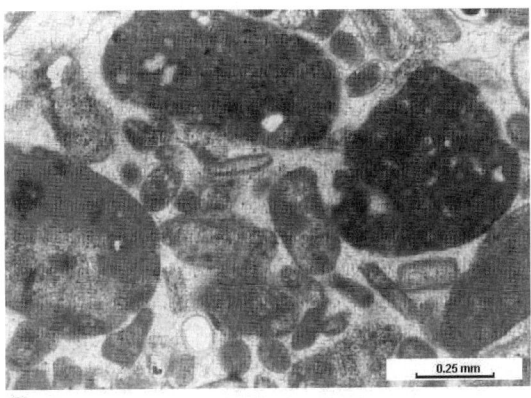

a

FIGURE   1.11 Intraclastic biosparite. This micrograph of the limestone Cordova Shell of Cretaceous age from Texas shows (a) intraclasts and (b) shell fragments. Intraclasts are fragments of limestone that were transported into the newly forming sediment. Large shells are pelecypod shells, many of which were dissolved, leaving large cavities called moldic pores.

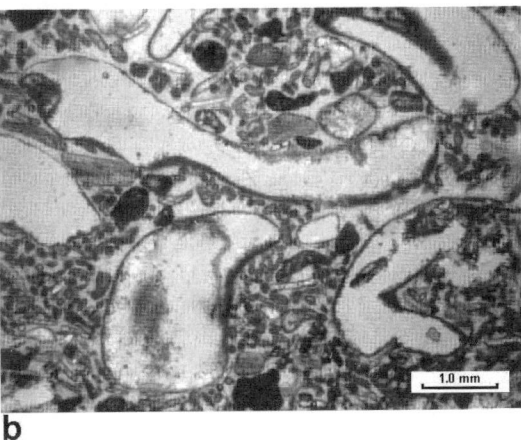

b

FIGURE 1.12 Oosparite. This micrograph of the St. Genevieve limestone of Mississippian age from Kentucky shows oolites, which are the main component of the rock. The oolites are cemented together by sparry calcite. Oolites show a well-developed concentric and radial structure, the former resulting from rolling carbonate particles in carbonate mud and the latter from the growth of algae encrusting particles as they grew.

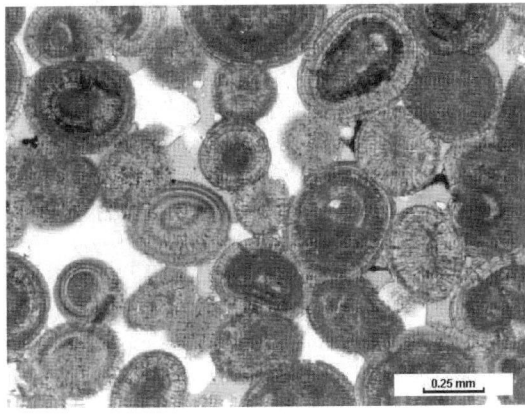

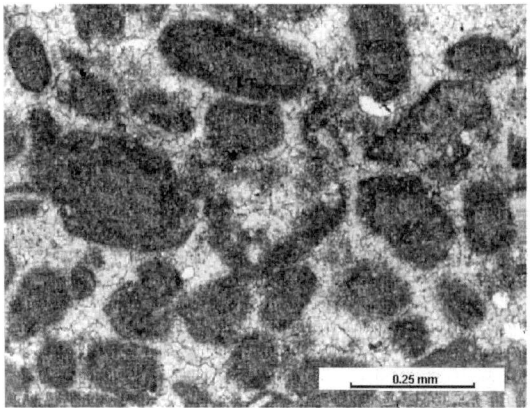

FIGURE 1.13   Pellsparite. This micrograph of Slade limestone of Mississippian age from Kentucky shows circular pellets that are excreta of bottom-living organisms. The pellets are cemented together by sparry calcite.

FIGURE 1.14   Sphinx limestone. Slabs of limestone from the Mokkatam formation of Eocene age are shown from which the Sphinx was carved. The sample on the left is sparse biomicrite (Fig. 1.6) and the one on the right is packed biomicrite (Fig. 1.7). We have used these limestone types in studies on durability reported in Chapters 8 and 9.

FIGURE 1.15   Slabs of the limestone types (from left to right): Cordova Cream (Fig. 1.9), Indiana (Fig. 1.8), Leuders (Fig. 1.10), and Cordova Shell (Fig. 1.11). We have used these limestone types in studies on reaction rates reported in Chapters 5 and 6.

TABLE 1.2   Classification of Limestone Based on Textural Components: Micrite, Allochems, and Sparry Calcite

Type I		Type II		Type III
		Mostly Micrite, Calcite Spar Subordinate or Lacking		Mostly Allochem, Calcite Spar Present
Micrite Only	Allochem Type	Micrite: Allochem > 0.5	Micrite: Allochem < 0.5	
Microcrystalline limestone, or lithographic limestone	Shell fragments Intraclast Ooids Pellets	*Sparse biomicrite Intramicrite Oomicrite Pellmicrite*	*Packed biomicrite*	*Biosparite Intrasparite Oosparite Pellsparite*

Source: Modified from Folk, R. L., Spectral subdivision of limestone types, in *Classification of Carbonate Rock*, Amer. Assoc. Petroleum Geol., Mem. Series, No. 1, 76, 1962.
[a]The rock type is given in italics. In type II limestone the ratio of micrite to allochem is given as more or less than 0.5.

1.4.4   Properties of Limestone

Limestone has many lithologic (mineral composition, texture, etc.) and structural properties, but our emphasis is on those that influence durability in outdoor exposure. These properties include porosity, stratification, and presence of stylolites.

1.4.4.1 Porosity Porosity is the volume of voids in a rock. Limestones range in porosity from being practically nonporous to having a porosity of 20% or higher. Whereas measurement of porosity and its features controlling durability are discussed in Chapters 8 and 9, here we will describe the origin of porosity.

The depositional texture of the sediment and the diagenetic processes control the porosity in limestone. For example, in micrite mud initial porosity is very large due to the abundance of fine-grained particles. This porosity is reduced by compaction as new sediment is laid down upon the old. In allochemical sediment, initial porosity is constructional, meaning that the space results from particles settling from water as best as they could fit. This porosity will be reduced during diagenesis by precipitation of cement in the pores. Along with these porosity-reducing processes are others that enhance them. When micritic material predominates in limestone, the passage of fresh underground water through pores may dissolve calcium carbonate from the matrix. In allochemical sediment, a similar change in porosity will occur when high-magnesian or aragonitic shells dissolve, creating *moldic porosity*. Also, algal encrustation and boring organisms may puncture holes in shells, connecting the skeletal pores with the constructional pores. The result is that limestone acquires an ink-bottle system of pores (Fig. 1.16), which is a major controlling factor of durability. The development of porosity in limestone is shown in Figure 1.17.

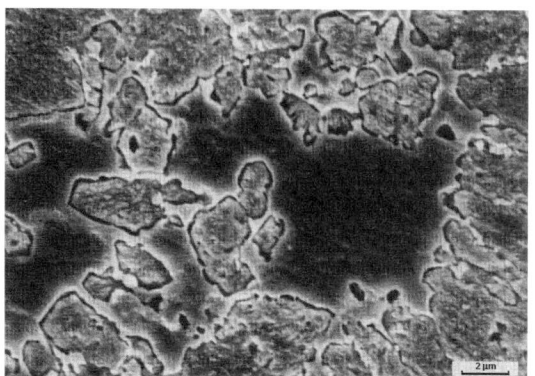

FIGURE 1.16 Sphinx limestone. This scanning electron micrograph shows ink-bottle pores, that is, large pores or bellies connected to each other through narrow channels or necks. Ink-bottle pores characterize most limestone types. *Source*: Chowdhury, A. N., Punuru, A. R., and Gauri, K. L., Weathering of limestone beds at the Great Sphinx, *Environ. Geol. Water Sci.*, **15** (3): 217–223, 1990.

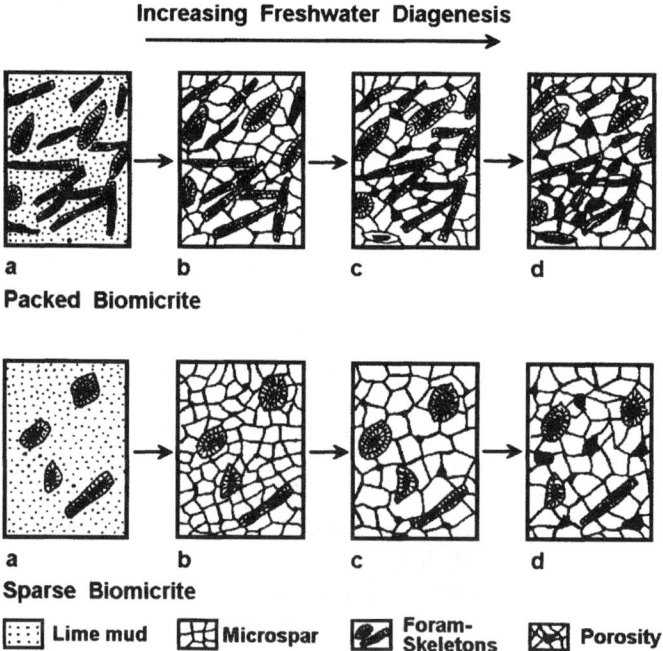

FIGURE 1.17 Evolution of porosity in limestone. Shown are Sphinx packed and sparse biomicrite: (a) primary depositional fabric; (b)–(d) progressive recrystallization and solution producing ink-bottle pores. *Source*: Chowdhury, A. N., Punuru, A. R., and Gauri, K. L., Weathering of limestone beds at the Great Sphinx, *Environ. Geol. Water Sci.*, **15** (3): 217–223, 1990.

1.4.4.2 Stratification Carbonate sediments accumulate grain by grain on the sea floor over a long period; the particle size varies with time. Thus, distinct layers or strata, each of uniform particle size, are produced.

The settling of particles from calm seawater forms fine-grained carbonate sediments. Stratification in limestone resulting from such sedimentation consists mostly of parallel beds. The coarse-grained sediment, however, is deposited from turbulent water, which, given a delicate balance of current velocity and particle size, produces cross-bedding. Cross-bedding refers to strata that are inclined with respect to the thicker, parallel-bedded strata in which they occur. Figures 1.18(a) and 1.18(b) show parallel bedding and cross-bedding, respectively.

Stratification influences durability. For example, blocks of dimensional stone placed in an ashlar such that the traces of bedding planes are exposed will weather differentially, thus producing a rough surface. This is why it is preferred that blocks are cut from thickly bedded strata and that the bedding surface itself is exposed to the outdoor environment.

a

b

FIGURE 1.18 Stratification in limestone. Like all sedimentary rocks, limestone shows parallel bedding developed on a smaller scale in a fine-grained limestone (a). In coarse-grained limestone, however, between thicker parallel-bedded strata sometimes cross-bedding (b) develops in which the layers are inclined with respect to horizontal bedding planes.

1.4.4.3 Stylolites Stylolites are structures produced by dissolution of calcite along bedding planes. In an outcrop in which traces of the stylolite surfaces can be

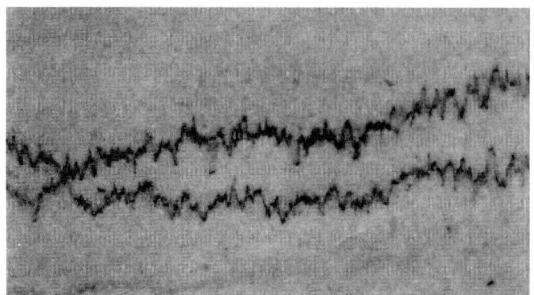

FIGURE 1.19 Stylolites, or crowfoot structures. Stylolites appear like sutures having formed subparallel to bedding planes by differential solution of limestone. Two layers of stylolites are shown with a dark coating of clay minerals. Limestone blocks with stylolites are often used only for indoor panels, because in outdoor exposure water may penetrate into the rock and cause deterioration.

seen, they look like a seismographic recording with the highs and lows being almost normal to the horizontal. Stylolite may thus be viewed as a seam where the upper and the lower bed interpenetrate in a jigsaw fit (Fig. 1.19).

Stylolites are considered to form by pressure solution. It is believed that inhomogenity in the strata causes unequal vertical pressures upon adjacent spots above and below the stylolite surface. Thus columns that remain in one layer fit into the conical depression in the other layer. A stylolite surface often has a coating of clay material that is believed to be the insoluble residue left after the calcium carbonate had been dissolved and removed from the site.

Slabs of limestone with stylolites are often used as indoor panels for the beauty they impart to polished surfaces. Often, they are also used in outdoor architecture, which is a poor choice due to the ease with which water can migrate into the rock and, on freezing, disrupt the stone.

1.5 DOLOSTONE

Dolostone is composed of calcium magnesium carbonate [$CaMg(CO_3)_2$] in the form of the mineral dolomite. The term *dolomite* is also often used as the rock name instead of dolostone.

1.5.1 Composition and Origin

The mineral dolomite forms primarily by the diagenetic replacement of calcite in limestone. Thus, limestone is the parent rock of dolostone, even though some primary recent precipitation of dolomite is known and some ancient dolostones are believed to have formed from original dolomite precipitation.

The replacement of calcite by dolomite is attributed to a solution–precipitation reaction. In this reaction calcite dissolves in the circulating water with a high Mg:Ca ratio, and simultaneous precipitation of dolomite takes place. Therefore, a large surface-to-volume ratio of a micrite carbonate sediment makes this sediment a

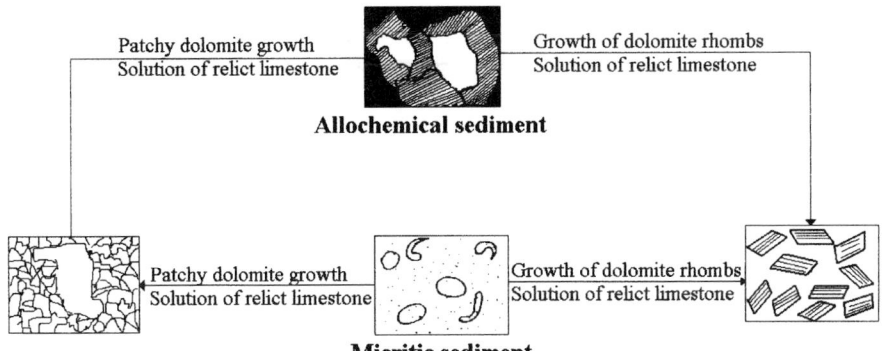

Patchy dolomite growth
Solution of relict limestone

Growth of dolomite rhombs
Solution of relict limestone

Allochemical sediment

Patchy dolomite growth
Solution of relict limestone

Growth of dolomite rhombs
Solution of relict limestone

Micritic sediment

FIGURE 1.20 Origin of dolostone. Dolostone forms by the replacement of calcite by dolomite. The formation of dolostone from lime mud or micritic limestone and sparite or framework limestone is shown, though micrite is more likely to change. (Modified from Murray, R. C., *J. Sediment. Petrol.*, **30:** 59, 1960).

common source of dolomite. Nevertheless, dolomitization of allochemical rocks is also known. Fig. 1.20 shows schematically the formation of dolostone from these two parent limestone types.

By definition, dolostone contains more than 50% dolomite. The term *calcareous dolostone* is often used when the rock contains 50% to 90% dolomite. A pure dolomite is considered to contain more than 90% dolomite. In a stoichiometrically balanced dolomite the ratio of calcium to magnesium is nearly 1:1. The X-ray diffraction peaks of dolomite are shown in Figure 1.21, which reveal a shift in the highest intensity peak position with a variable Mg:Ca ratio. Thus, an X-ray diffraction trace can be used to obtain an estimate of the magnitude of replacement of Ca by Mg.

1.5.2 Classification

Dolostone is made of variable-size rhombic grains of dolomite with planar or nonplanar surfaces. Formation of these grains, as in any other crystallization process, is controlled by the rate of nucleation and the growth kinetics (Sibley and Gregg, 1987).

Dolomitization is believed to occur from solution at elevated temperature in variable super-saturation concentrations. The kinetics (rate of growth) of crystal growth from solution suggests that planar surfaces are preferred at lower supersaturation and low temperature, whereas nonplanar crystal faces are indicative of higher supersaturation and higher temperatures. The crystal size, however, is controlled by the interaction between nucleation rate and growth kinetics. If the nucleation rate increases faster than the growth rate, many small crystals will form; if this condition is reversed, a small number of large crystals will occur in a rock. The occurrence of a bi- or polymodal grain-size distribution in a dolostone, therefore, suggests a changing environment during the replacement–reaction history of the rock.

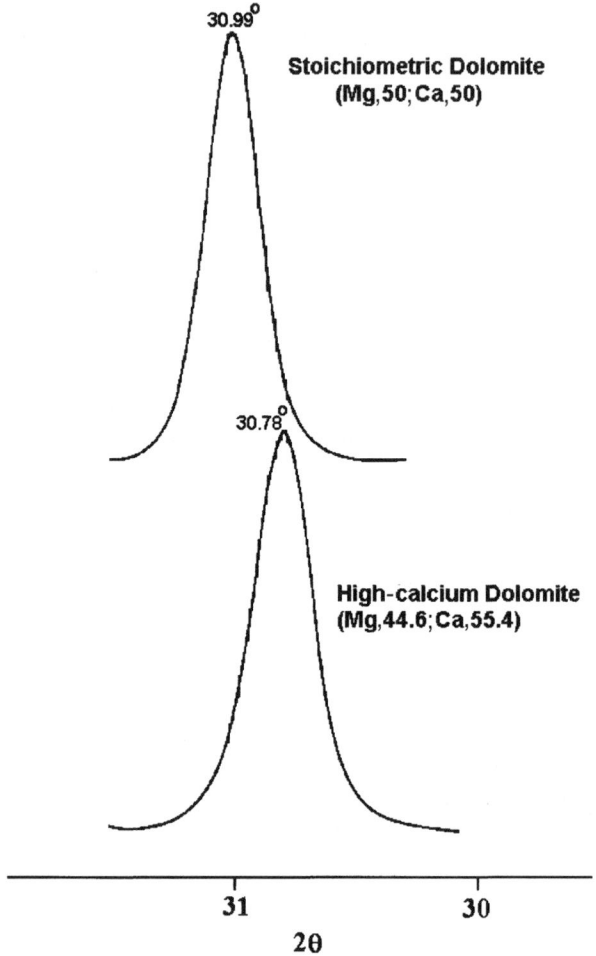

FIGURE 1.21   X-ray diffraction traces of calcareous dolostone. A typical XRD highest intensity peak is shown for a stoichiometrically balanced dolomite (top), and the shift in the peak for dolomite with lower Mg:Ca ratio is presented (bottom).

Change in porosity of the parent limestone and that of the dolomitic limestone as the dolomitization progresses add another dimension to the growing dolomite grains. When adequate pore space is available, the crystals will be euhedral, that is, with well-defined forms. Otherwise, the crystal faces may grow into each other, making a sub- or anhedral crystal.

Thus, a dolomite rock can be stoichiometrically balanced with respect to the Ca:Mg ratio; it may be unimodal or polymodal, planar or nonplanar, and euhedral or subhedral. Common dolostone types are assigned names after these parameters. Combining texture, porosity, and chemical composition, the Laurel dolomite often used in studies given in this book is described as a stoichiometric, polymodal, planar-e

FIGURE 1.22 These slabs of dolostone are from two rock units in the Louisville area: the Big Blue dolomite (a) occurs as a highly receded bed in outcrops of Louisville limestone (Fig. 1.26); the Laurel dolomite (b) forms prominent bluffs in the area where exposed. The Louisville limestone and the Laurel dolomite are formations of Silurian age. We used Laurel dolomite in studies on reaction rates reported in Chapters 5 and 6.

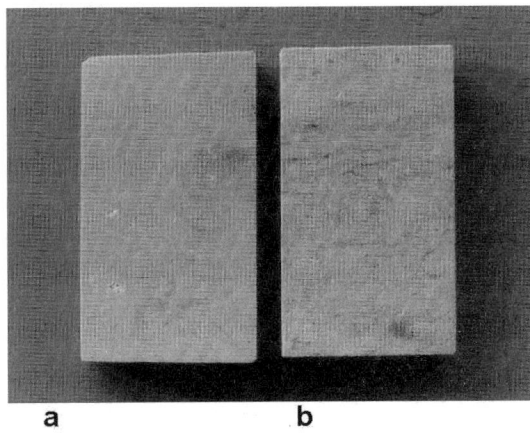

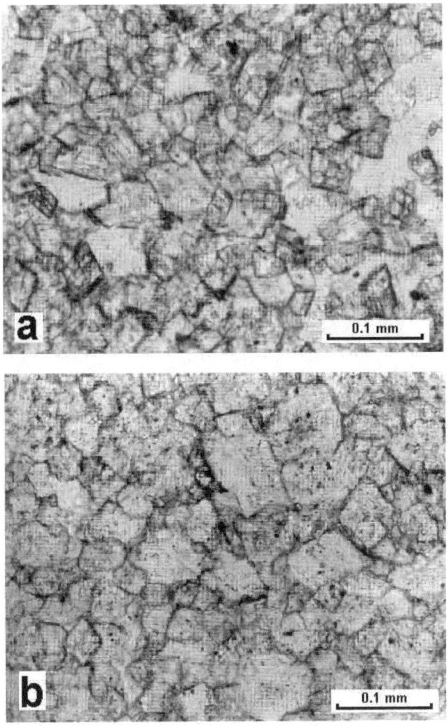

FIGURE 1.23 Dolomite crystals. A micrograph of sample of (a) Big Blue dolomite and (b) Laurel dolomite shows typical rhombic form of the crystals. However, the grains are smaller in the Big Blue than in the Laurel dolomite.

(euhedral) dolostone. Figures 1.22 and 1.23 show *Laurel dolomite* and a fine-grained dolostone.

Further description of a dolostone may include reference to the textural features of the parent limestone. Even though the original structure and shape are often completely obliterated in most dolostones, some dolostone varieties may selectively preserve the original attributes. Several terms are used to describe the original allochem. *Pseudomorphs* contain the original calcite structure, but the mineral is now dolomite. *Ghosts*, which are inclusions, vaguely demarcate the original outline of the structure; an example is a fossil shell. A *mimic replacement* is the preservation of the form and internal structure of the allochem.

1.5.3 Porosity

Porosity and pore structure of dolostone differ greatly from those of the parent limestone. A well-dolomitized rock, such as the Laurel dolomite mentioned earlier, exhibits the typical pore shape (Fig. 1.24) between the rhombic dolomite crystals. In dolostone porosity is higher than that of the parent limestone because of the 13% volume shrinkage of dolomite compared with calcite based on mole for mole replacement of larger Ca^{2+} by Mg^{2+} ions as shown in the following equation:

$$Mg^{2+} + 2CaCO_3 \rightarrow MgCa(CO_3)_2 + Ca^{2+} \tag{1.2}$$

Another factor controlling porosity change is the nature of the parent limestone. For example, in the dolomitic rocks of the Mississippian Charles Formation in the Midale Field, Saskatchewan, a 50 ft thick sequence exhibits all degrees of dolomitization. Murray (1960) reported that in these calcareous dolomites, where the replacement is not advanced, porosity seems to reduce at first apparently due to the compaction of the parent micritic sediment. The dolomitization, however, continued

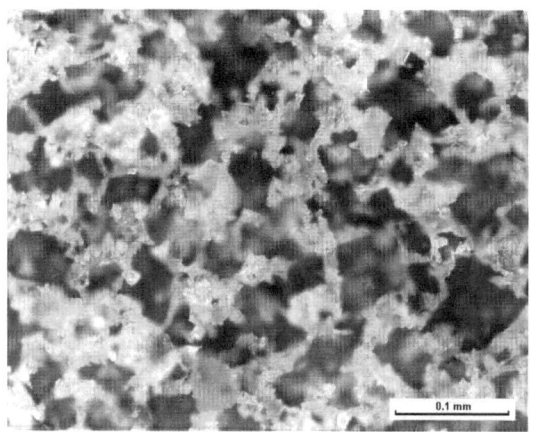

0.1 mm

FIGURE 1.24 Pores in dolostone. A micrograph of a thin section of Laurel dolomite is shown, which, before final polish, was impregnated with epoxy. After the polish, the sample was deeply etched with hydrochloric acid to dissolve the surface carbonate. The bright configurations at the micrograph surface are the epoxy fillings representing the pore structure. Note that the pores roughly correspond with the crystal outline shown clearly in Figure 1.23(b), but they also are visible as dark areas here.

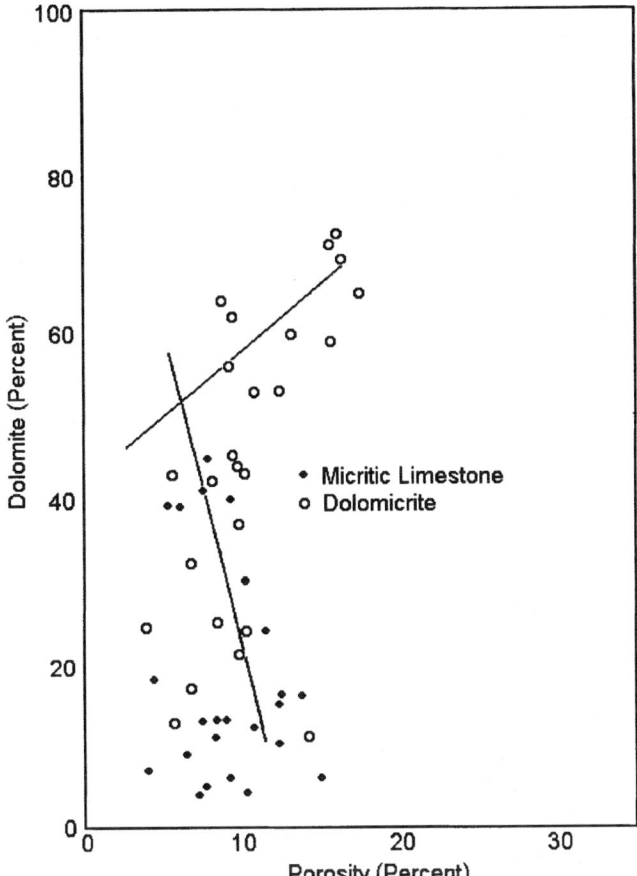

FIGURE 1.25 Pore-volume change resulting from dolomitization. A change in the volume of pores during increasing diagenesis of a micritic carbonate sediment is shown. At first a reduction in porosity occurs due to compaction. In advanced diagenesis when the dolomitization is nearly complete, the pore volume increases due to the smaller size of magnesium atoms that replace the larger calcite atoms. (Modified from Murray, R. C., *J. Sediment. Petrol.*, **30:** 59, 1960).

after the compaction had ceased. The dolostones with advanced dolomitization do indeed show an increased porosity (Fig. 1.25).

Some dolomites have large, irregularly shaped holes or void spaces called vugs. The vugs are due to the dissolution of residual calcite, which, perhaps due to scarcity of circulating water caused by the low permeability of rock, could not be dolomitized. These vugs make dolomite excellent reservoirs for hydrocarbons (petroleum or natural gas) or sites for disposal or storage of fluids.

The knowledge of the nature of dolostone, the degree of dolomitization, and the porosity characteristics allow some insight into the potential behavior of dolomite when subject to weathering. The Laurel dolomite forms prominent bluffs in the

FIGURE 1.26 Outcrop of Louisville limestone. The receded Big Blue dolomite bed present between two limestone layers can be seen. The highly intense weathering of the Big Blue dolomite is largely due to a high microporosity.

Louisville area, whereas the Big Blue bed in the *Louisville limestone* is seen highly recessed in natural outcrops (Fig. 1.26). Figures 1.22(a) and 1.23(a) are samples taken from this outcrop.

1.6   MARBLE

1.6.1   Composition and Origin

In the geological sense, marble is a metamorphic rock made of the mineral calcite or dolomite. Commercially, however, other crystalline rocks that can be well polished and used in architecture are also called marble. These rocks include Verde Antique, which is made largely of the silicate mineral serpentine but is traversed by calcite veins, and alabaster, which is often considered to be composed of gypsum, $CaSO_4 \cdot 2H_2O$. Also, certain fossiliferous limestones, which may be partially or not at all metamorphosed, are termed marble because their fossil content enhances their beauty. A well-known "marble" of this type is the American Tennessee marble, some of which was used in the Lincoln Memorial.

The parent rock of marble is limestone. The process of metamorphism involves the exposure of limestone to elevated temperature and pressure brought about by deep burial followed by intense tectonic activity. This is why the occurrence of marble nearly coincides with the fold-mountain belts. For example, the American Vermont, Georgia, and Carthage marbles are associated with the Appalachian and the Ozark belts, and the Yule marble, occurring in Colorado, is associated with the Rocky Mountains.

As a result of intense metamorphism, some calcite grains in the limestone are dissolved and reprecipitated around other grains, making the grains in marble larger than those in the parent limestone. The recrystallization causes the disappearance of the original structure so that the fossil shells disappear, the porosity is obliterated as the growing crystals interlock, and the bedding planes vanish as the beds become fused. Thus, marble is a highly compact, nearly nonporous rock made of large crystals of calcite and devoid of any parent structure and porosity. Nevertheless, even one of the most famous marble varieties, the Carrara marble, may have recognizable relicts of original fossils.

Since many limestone types are impure rocks containing minor quantities of clay minerals, quartz, and bituminous matter, marble inherits these impurities, which also can become altered in the process of metamorphism. For instance, the clay minerals may be converted to mica; the Georgia marble has pockets of phologopite, a mica mineral. The bituminous matter in limestone becomes graphite in marble and is often responsible for black streaks. Quartz, depending upon the degree of metamorphism, reacts with calcite by the reaction

$$CaCO_3 + SiO_2 \rightarrow CaSiO_3 + CO_2 \qquad (1.3)$$

producing wollastonite ($CaSiO_3$), which is grayish, yellowish, or greenish and has metalloid luster. Dolomite in limestone can convert to several other colored ferromagnesian silicate minerals of the olivine and pyroxene group (see Sec. 2.2). Weathering of these minerals, in turn, produces other colored minerals such as talc and serpentine, which in their own right make independent varieties of marble. Thus marbles comes in various colors. The pure white variety is the one most cherished by sculptors. This variety forms from limestone that lacks any occluded impurities.

Different types of marble are recognized on the basis of petrofabrics, which includes orientation, size distribution, and the packing of grains. The following descriptions are of the common varieties of American marble, but includes Carrara marble, which is used worldwide in sculpture and cemetery monuments.

1.6.2 Some Marble Varieties

1.6.2.1 Georgia Marble. Georgia marble, quarried in Georgia in the United States, appears in various shades of pink (Cherokee type) and gray. Slabs of Georgia marble appear mottled due to the presence of the mineral phologopite when it is concentrated in bands.

This marble has a coarse-grained fabric [Fig. 1.27(a)]. The grain size ranges between 700 μm and 4,000 μm with the mean grain size being 1,300 μm. The grains have irregular shapes and are interlocked into a tightly packed mass.

1.6.2.2 Alabama and Vermont Marbles. These marble types are pure white even though in large slabs streaks of graphite lend a grayish appearance locally. The texture is fine-grained: grains range in diameter from 20 μm to 300 μm and 70 μm to 800 μm, respectively, while the mean grain sizes are 140 μm and 220 μm. The grains [Fig. 1.27(b)] have irregular shapes and are interlocked tightly but not as tightly as in the Georgia marble.

1.6.2.3 Carrara Marble. Carrara marble is mostly pure white and has a sugary texture due to nearly equidimensional, rounded calcite grains [Fig. 1.27(c)], giving it

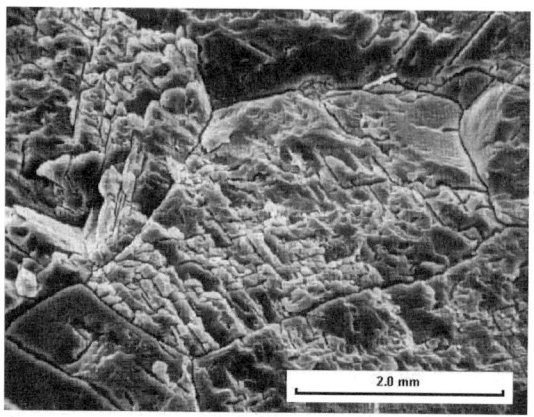

a

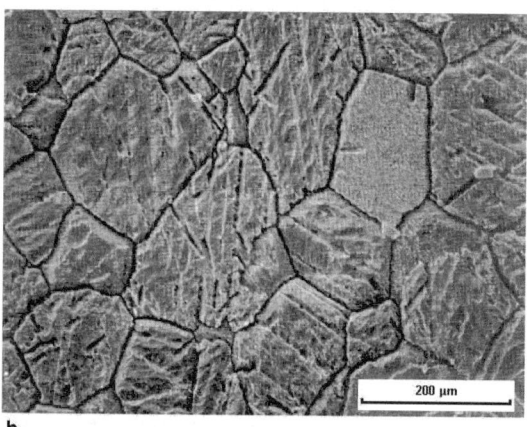

b

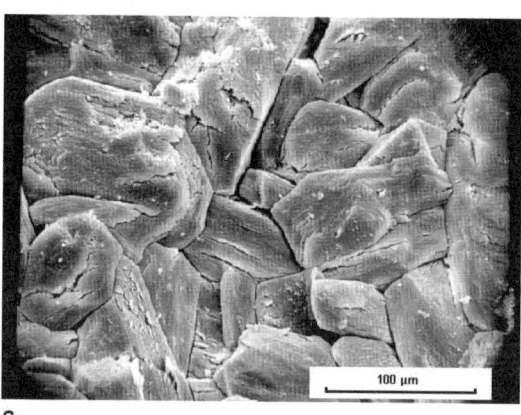

c

FIGURE 1.27 Marble types. Scanning electron micrographs are presented of different types of marble distinguished by grain size. (a) The grain of Georgia marble in the middle is nearly 2 mm in diameter and the rhombic intersecting lines in the grain represent the typical calcite cleavage planes. (b) Vermont marble is relatively fine-grained. (c) Cararra marble is fine-grained and has a sugary texture. We have used these marble types in studies on weathering reported in Chapters 5 and 6.

somewhat less compacted fabric. The grain size ranges from 50 μm to 200 μm, with a mean diameter of 120 μm.

Because of the lack of pores marble is quite durable in physical weathering, if properly quarried (Sec. 4.3.3). However, chemical reaction rates vary depending upon the fabric (Sec. 5.3.1).

1.7 TRAVERTINE

Travertine, also called travertine marble in the stone industry, is a freshwater limestone formed on land from hot springs. Travertine is commonly a variegated, highly porous rock, often with finely laminated structure (Fig. 1.28). The name *travertine* originates from the locality of Travertino, the Latin name of modern Tivoli, near Rome, where deposits hundreds of meter thick occur. Here, travertine is known to have been quarried for more than 2,000 years and is used in such well-known structures as the Coliseum of Rome. Even the footpath pavements and fountains in the city of Rome are made largely of travertine. In the United States, among many other edifices the newly constructed Getty Center in Los Angeles is made of travertine dimensional blocks imported from Italy.

Travertine forms from the freshwater of natural springs and small rivers by mostly biochemical processes in much the same fashion as the marine carbonates. Common in the geological setting of the travertine near Rome and that of western United States in Wyoming and Idaho is the proximity of these deposits to volcanically active areas that generate hot, CO_2-enriched subsurface water in the limestone country rock whence the calcium bicarbonate, $Ca(H_2CO_3)_2$, is mobilized by solution. These hot spring waters, averaging nearly 20°C, are supersaturated with respect to calcium (Ca^{2+}) and bicarbonate [$(HCO_3)^-$] ions. Near the vents, where the temperature is much higher, dissolved CO_2 from the earth's interior quickly escapes to equilibrate with the atmospheric partial pressure of CO_2. This results in the precipitation of rather white calcium carbonate. Also, in this proximity, bacteria that favor elevated temperature for their existence make internal rodlike structures and external botryoidal (like bunch of grapes) forms of calcite.

Extensive travertine deposits, however, form farther from the vents—the points of origin of streams—where shallow water flows in sheets or cascading streams and

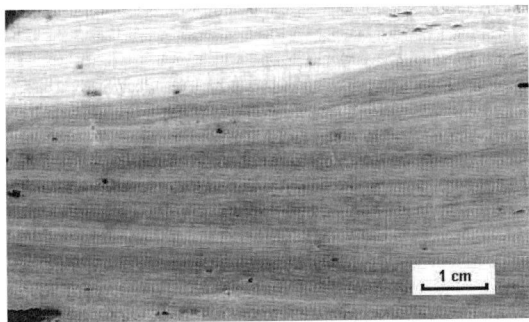

FIGURE 1.28 Travertine. A cross-section of a slab of travertine from Tivoli, Italy, is shown, with fine lamination that is due to the seasonal rhythmic growth supported by bacteria.

allows growth of algae and mosses at the rock floor. The photosynthesis by these organisms as explained in the discussion on marine carbonates aids $CaCO_3$ deposition.

Travertine is highly porous, even vuggy or cavernous, laminated rock. The high porosity is due to the degradation of the organic matter, the plant life that is actually responsible for the deposition of travertine. However, some of the pores that are large cavities may be filled with large mosaic calcite of secondary origin. The lamination is mainly due to seasonal rhythmic growth; the spring laminae are darker and thinner. Common colors seen even in a single slab vary from yellow to tan and brown due to contamination with iron oxides and hydrates.

Travertine is an ornamental stone of choice in modern common non-load-bearing, high structures in which the weight of the masonry wall, or curtain wall, should be as low as possible. For such external veneers (Fig. 1.29) as well as for pavements, the surface of travertine panels, before final polish, is often impregnated with some polymeric slurry to provide the stone resistance to weathering. Massive, roughly hewn blocks are also used in load-bearing construction.

FIGURE 1.29   Travertine. Slabs of travertine used in external cladding of a building in downtown Toronto, Ontario, are shown.

SUMMARY

Carbonate rocks consist of limestone, dolostone, and marble and are made of the minerals calcite and dolomite. Methods of identification of these rocks are given.

Limestones form by the diagenesis of carbonate sediments, which are laid down upon shallow sea floor, most commonly by biochemical activity. Limestones, like carbonate sediments, are classified on the basis of texture, which consists of a microcrystalline matrix, allochemical framework grains, and calcite spar cement. Major classes of limestone are micrites and sparites.

A property related to texture is the porosity. Although all limestone types are characterized by ink-bottle pores, the micrites are, in general, more microporous than sparites. We will show later (Chap. 9) that micrites, due to their high microporosity, are less durable. Nevertheless, the magnitude of durability varies within the micrites and sparites based upon the relative proportion of micro- and macropores.

Dolostone is made of dolomite, which forms by the partial replacement of calcium in calcite by magnesium. Dolostones are classified on the basis of the degree of this replacement and the size and shape of crystals.

Marble is a metamorphic rock, the types of which are distinguished on the basis of particle size and the degree of packing. Georgia marble, a coarse-grained, highly compact marble, and Vermont and Carrara marble, which are fine-grained, are described.

Finally, travertine, a freshwater limestone that forms from hot springs, is described.

REFERENCES

Dunham, R. J., Classification of carbonate rocks according to depositional texture, in *Classification of Carbonate Rocks*, Amer. Assoc. Petroleum Geol., Mem. Ser., No. **1**, 62–84, 1962.

Folk, R. L., Spectral subdivision of limestone types, in *Classification of Carbonate Rocks*, Amer. Assoc. Petroleum Geol., Mem. Ser., No. **1**, 108–121, 1962.

Folk, R. L., Nannobacteria: surely not figments, but what under heaven are they?, *Natural Sci.*, **1** (3): 1–10, 1997.

Murray, R. C., *J. Sediment. Petrol.*, **30**: 59, 1960.

Sibley, D. F., and Gregg, J. M., Classification of dolomite rock textures, *J. Sediment. Petrol.*, **57**: 967–975, 1987.

CHAPTER 2

Noncarbonate Minerals in Carbonate Rocks

2.1 INTRODUCTION

Carbonate sediments, the parent of sedimentary rocks, are deposited in segregated marine basins without much flux from outside. Also, the low solubility of calcium carbonate precludes simultaneous precipitation of other evaporite minerals that are more soluble. One would therefore expect limestone to be devoid of noncarbonate matter. Nevertheless, limestone often contains clastic particles (clay and silt) and water-soluble salts (commonly gypsum and halite). Clastic particles are defined as particles that have been moved individually from the site of their origin as opposed to salts that form as a result of chemical precipitation from ions by the evaporation of seawater. These noncarbonate minerals influence the durability of limestone, discussed in Chapter 3. In this chapter we will give the origin of these minerals, methods to concentrate them from the rock mass in which they are sparsely distributed, and their identification by the commonly applied techniques of X-ray diffraction.

2.2 CLAY MINERALS

All minerals can be assigned to five or six groups based on their anionic composition, which is controlled by the environment in which the minerals crystallize. Clays are silicate minerals defined by the anion $(SiO_4)^{4-}$. They are chemically distinct from carbonate minerals, such as calcite $(CaCO_3)$, in which the defining anion is carbonate $[(CO_3)^{2-}]$. Thus there is no genetic association between the two that would warrant their coexistence. Nonetheless, clay minerals and other clastic particles are often found occluded in carbonate rocks. Derived from land sources, these particles have been washed into the marine environment of carbonate sediments. When present, clay minerals are a major cause of stone decay. In the following, we will give the origin, the structural chemistry, and the X-ray diffraction method for the identification of these minerals.

2.2.1 Origin

The chemical weathering of primary silicate minerals that crystallize from lava or magma produces clay minerals. To facilitate understanding of the origin of clay minerals we will first describe the origin of the parent silicate minerals. The sequence of crystallization of minerals and the silicate structures are shown in Figures 2.1 and 2.2.

Silicate minerals crystallize from lava and magma, which are silicate melts made largely of the cations of potassium (K^+), sodium (Na^+), magnesium (Mg^{2+}), calcium (Ca^{2+}), iron ($Fe^{2+,3+}$), aluminum (Al^{3+}), silicon (Si^{4+}), and the anion oxygen (O^{2-}). As the melt cools to below nearly 1200°C, Si^{4+} and O^{2-} first combine to form the complex silicate ion or radical $(SiO_4)^{4-}$ modeled as a tetrahedron: a pyramidal form in which each side is a triangle. In a tetrahedron, the silicon ion can be seen as occupying the center while the oxygen ions occupy the corners. The negative charge on the $(SiO_4)^{4-}$ ion allows it to bond with cations in the melt to make electrically neutral minerals. In the simplest case, Fe^{2+} and Mg^{2+} bond with isolated $(SiO_4)^{4-}$ tetrahedra and form a group of minerals represented by olivine $(Mg,Fe)_2SiO_4$, which are the first to crystallize from magma. In further cooling, several silicon–oxygen tetrahedra link together by sharing oxygen and form a single chain structure. Single chains may join to form double chains, and then sheets can be formed when many chains merge in a plane. Most of these isolated tetrahedra, single chains, double chains, and sheets of polymerized $(SiO_4)^{4-}$ tetrahedra combine with appropriate cations to give rise to

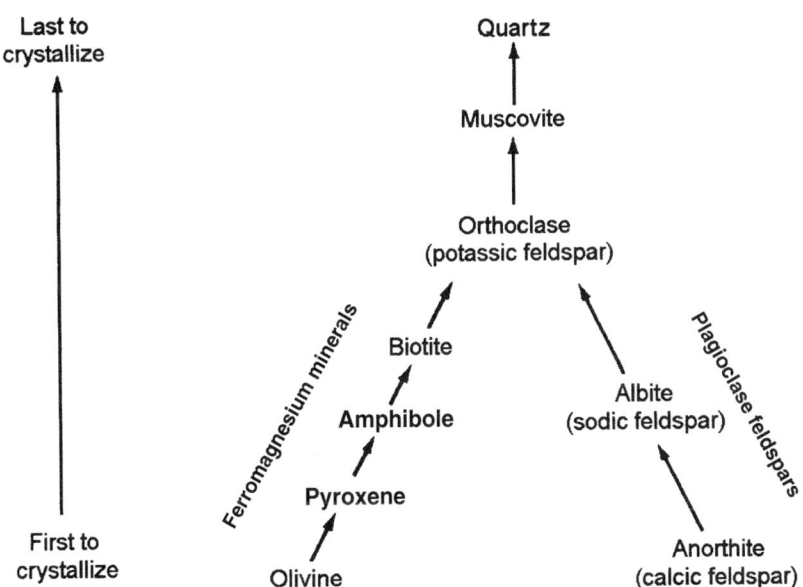

FIGURE 2.1 Primary silicate minerals. The sequence of crystallization of silicate minerals from cooling lava or magma is shown.

Polymerization of SiO_4 tetrahedra	Structure	Mineral	Formula
Isolated tetrahedra		Olivine	$(Mg,Fe)_2SiO_4$
Single-chain structure	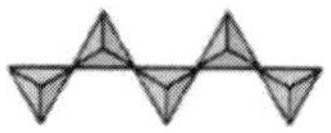	Pyroxene	$(Mg,Fe)SiO_3$
Double-chain structure	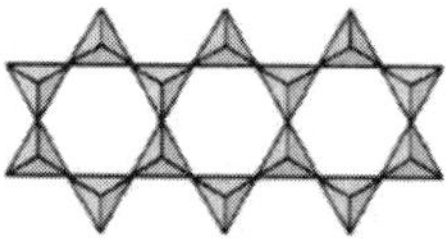	Amphibole	$(Ca_2Mg_5)(Si_8O_{22})(OH)_2$
Sheet structure		Biotite	$K(Mg,Fe)_3(AlSi_3O_{10})(OH)_2$
		Kaolinite	$Al_2Si_2O_5(OH)_4$
Framework structure		Feldspar (orthoclase)	$KAlSi_3O_8$
		Quartz	SiO_2

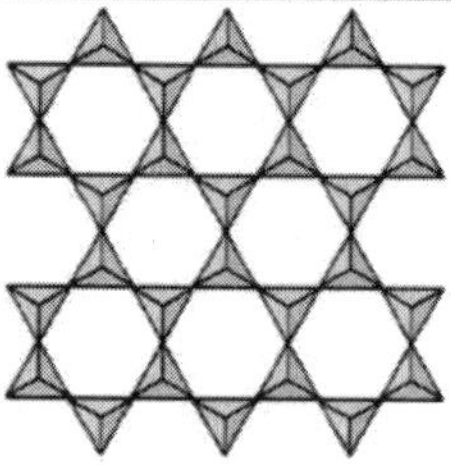

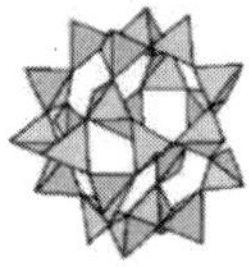

FIGURE 2.2   Structure of primary silicate minerals. The fundamental unit of silicate structure is $(SiO_4)^{4-}$ radical, viewed as a tetrahedron. Silicate structures form by combining isolated tetrahedra or by their polymerization. Cations attach to these structures, making various groups of the silicate minerals.

various groups of silicate minerals. The remaining melt contains $(SiO_4)^{4-}$ tetrahedra only. These silicon tetrahedra join in a framework structure, giving rise to the mineral quartz (SiO_2), which is the last mineral to crystallize from magma at nearly 750°C.

Common primary silicate minerals involved in the formation of clay minerals are feldspars, which crystallize from the early to late stages of magmatic differentiation. Early to form are plagioclase feldspars and the last alkali feldspar, a common example of which is orthoclase ($KAlSi_3O_8$). In the following we will give chemical reactions describing weathering of orthoclase that leads to formation of clay minerals; note that other feldspars and some ferromagnesian minerals weather by similar reactions.

Clay minerals form largely from feldspars that weather by hydrolysis, a process in which water reacts with the mineral.

$$KAlSi_3O_8 + H_2O \rightarrow Al_2Si_2O_5(OH)_4 + SiO_2 + K^+ \qquad (2.1)$$

Silica (SiO_2) and potassium (K^+) are dissolved in water and are drained away. Kaolinite [$Al_2Si_2O_5(OH)_4$], a clay mineral, settles in place or is transported and deposited elsewhere. The dissolution of feldspar in pure water is a very slow process but is accelerated when the rain or the subsurface water dissolves carbon dioxide from the atmosphere or the soil, forming carbonic acid (H_2CO_3). These reactions can be expressed as follows:

$$CO_2 + H_2O \rightarrow H_2CO_3 \qquad (2.2)$$

$$2KAlSi_3O_8 + 2H_2CO_3 + H_2O \rightarrow Al_2Si_2O_5(OH)_4 + 4SiO_2 + 2K^+ + 2HCO_3^- \quad (2.3)$$

Clay minerals are hydrous aluminum silicates. However, in the process of weathering some aluminum (Al^{3+}) may be replaced by magnesium (Mg^{2+}) and aluminum may replace some silicon (Si^{4+}) in the mineral structure, a process called *isomorphous replacement*. Such substitutions and the manner in which atoms are arranged in the mineral structure create a large variety of clay minerals.

2.2.2 Classification

Clay minerals are phyllosilicates, that is, they have a layered structure (*phyllon*, leaf), meaning that the atoms are arranged in sheets and that the sheets are chemically bonded to each other and form layers.

Two types of sheets are recognized, namely, *tetrahedral* sheets and *octahedral* sheets (Fig. 2.3). The tetrahedral sheet is made of silicon–oxygen tetrahedra linked through oxygen, and the octahedral sheet consists of octahedra with each octahedron containing aluminum or magnesium surrounded by six oxygen and hydroxide anions. If aluminum is the cation, then the sheet is termed dioctahedral; if magnesium, then the sheet is trioctahedral. The octahedrons are linked by sharing oxygen and hydroxal ions. The tetrahedral and the octahedral sheets are connected to form layers by the unshared oxygen of the tetrahedral sheet and the divalent (Mg^{2+}) and trivalent (Al^{3+}) cations in the octahedral sheet.

Three main clay-mineral structures are recognized based on the number of sheets stacked in the basic structure, called the unit cell. These structures are (1) two-sheet structure, (2) three-sheet structure, and (3) three-plus-one sheet structure. Kaolinite, for instance, possesses two-sheet structure [Fig. 2.4(a)]. This means that a single flake, or layer, of kaolinite consists of one Si–O tetrahedral sheet and one Al–OH octahedral sheet. A three-sheet structure is one in which an octahedral sheet is sandwiched between two tetrahedral sheets [Fig. 2.4(b)]. In a three-plus-one sheet structure four alternating tetrahedral and octahedral sheets are present in one unit cell. After these structures the clay minerals are termed 1:1, 2:1, and 2:1:1 clays.

Further distinction among clay minerals is based upon the ionic population of the octahedral sheet. If trivalent aluminum ions (Al^{3+}) are present in the centers of octahedra, only two-thirds of the octahedral centers are filled, termed a dioctahedral

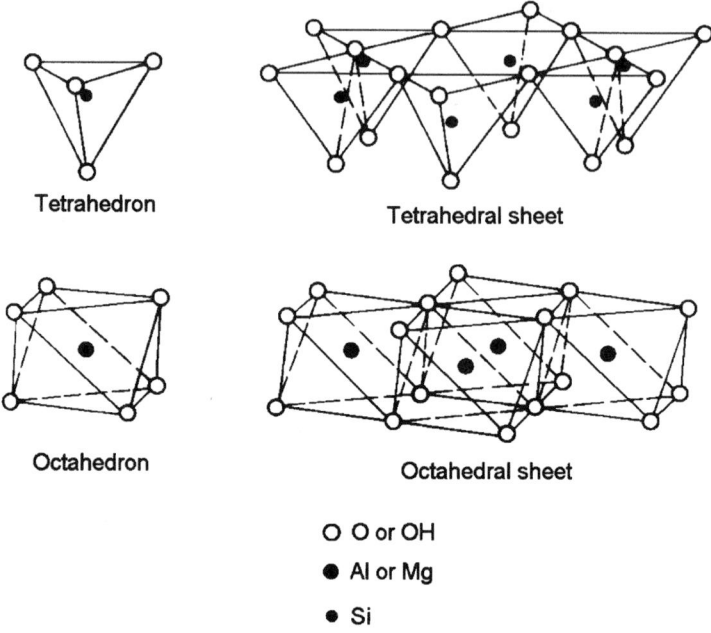

Tetrahedron

Tetrahedral sheet

Octahedron

Octahedral sheet

O   O or OH

● Al or Mg

• Si

FIGURE 2.3   Structure of clay minerals. Clay minerals have a layered structure formed by the stacking of tetrahedral and octahedral sheets. Shown are the structure of the basic tetrahedron and octahedron, their polymerization into sheets, and the site of cations. *Source*: Grim, R. E., *Clay Mineralogy*, 2nd ed., New York: McGraw-Hill, p. 52. Reproduced with permission of the McGraw-Hill Companies.

population or *gibbsite* sheet. However, if divalent magnesium ions (Mg^{2+}) are present, then all the centers are filled, making this a trioctahedral population or *brucite* sheet.

Finally, clay minerals are characterized by the presence or absence of bonded water between the layers. The clays with interlayer water are called swelling clays and those without are nonswelling clays. Water can enter certain clay structures because the tetrahedral and octahedral sheets between two adjacent layers are not tightly bonded due to the electrical imbalance. Table 2.1 shows commonly used classification of clay minerals. The basal *d spacing* characterizing each mineral is the distance between two consecutive layers and is measured by X-ray diffraction as given in the section on identification.

Kaolin Group.   The kaolin group consists of 1:1 clays, that is, each layer is made of two sheets only. In the case of the species *kaolinite*, which is nonswelling, the basal *d* spacing is 7 Å. *Halloysite* is a swelling species of the kaolin group; its *d* spacing is 7.2 Å and 10 Å for $2H_2O$ and $4H_2O$ interlamellar water.

Kaolinite is pure white clay, being exclusively silicoaluminous. Other species of the kaolin group are distinguished by the partial replacement of aluminum in the

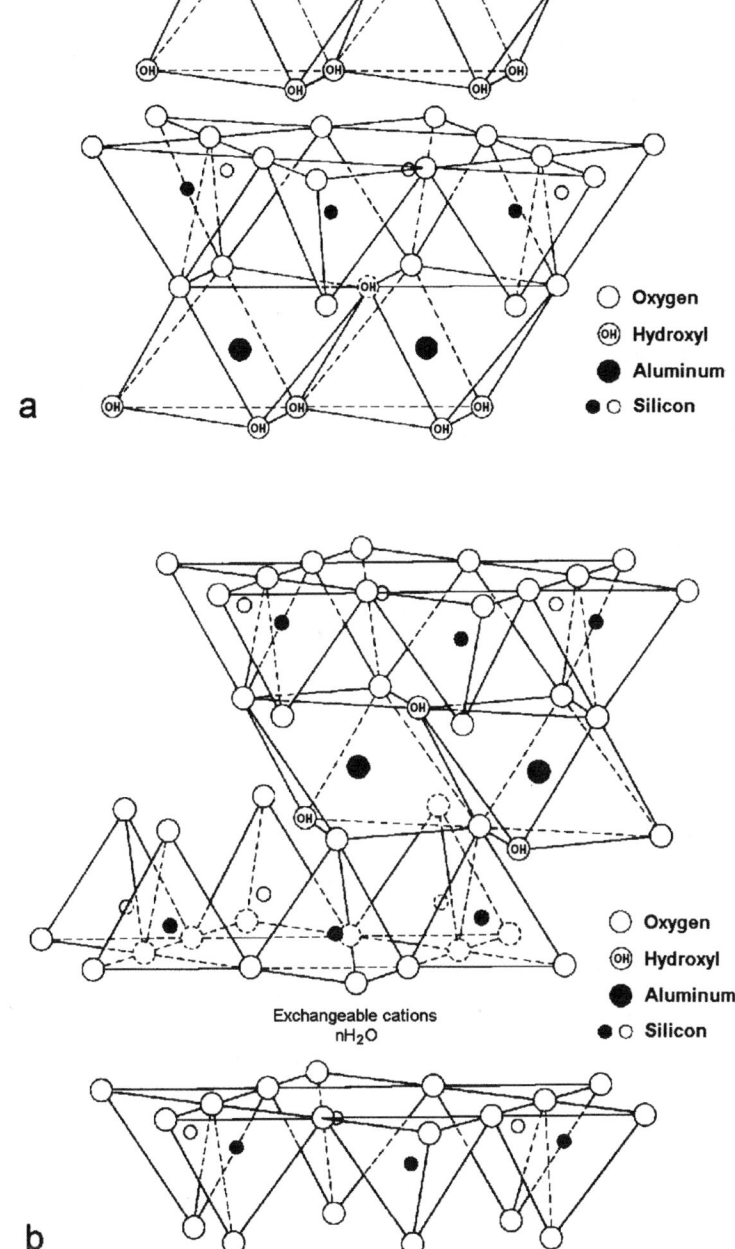

FIGURE 2.4 Unit cells of clay minerals. Clay minerals have two, three, or three plus one sheets in a unit cell. Clay minerals have a two-sheet structure (a) if the unit cell consists of one tetrahedral and one octahedral sheet, and a three-sheet structure when an octahedral sheet is sandwiched between two tetrahedral sheets (b). *Source*: Grim, R. E., *Clay Mineralogy*, 2nd ed., New York: McGraw-Hill, p. 52. Reproduced with permission of the McGraw-Hill Companies.

TABLE 2.1 Common Classification of Principal Clay Minerals

Species	Structure	Octahedral Sheets	Swelling
Kaolinite $Al_4Si_4O_{10}(OH)_8$	1:1	Dioctahedral	Nonswelling
Halloysite $Al_4Si_4O_{10}(OH)_8 \cdot 4H_2O$			Swelling
Septechlorite $Mg_6Si_4O_{10}(OH)_8$		Trioctahedral	Nonswelling
Illite $K_{0-2}(Al,Si)_8O_{20}(OH)_4$	2:1	Dioctahedral	Nonswelling
Montmorillonite $Al_4Si_8O_{20}(OH)_4 \cdot nH_2O$			Swelling
Vermiculite $Mg_6Si_8O_{20}(OH)_4 \cdot nH_2O$		Trioctahedral	Swelling
Chlorite $Mg_3Si_4O_{10}(OH)_2 \cdot Mg_3(OH)_6$	2:1:1	Trioctahedral	Nonswelling

octahedral sheet by iron, magnesium, nickel, or manganese, which imparts various colors to these forms.

Illite Group. Illites have a 2:1 structure. The most common species of this group is *illite*, which is nonswelling. The d spacing in *illite* is 10 Å. In addition to structure, the characteristic of this group is the presence of potassium ions in the interlamellar space. Partial replacement of aluminum by iron in the octahedral sheet results in the formation of *glauconite*, a green clay mineral of marine origin.

Montmorillonite Group. *Montmorillonite*, also called *smectite*, is a swelling clay with 2:1 structure. Its basal d spacing varies from 10 Å to 17.5 Å. Whereas *montmorillonite* is the most common representative of this group, *beidelite* is one in which silicon is partially replaced by aluminum. In *nontronite* and *seponite*, aluminum is partially replaced by iron and magnesium, respectively.

Vermiculite Group. This group of minerals consists of trioctahedral 2:1 swelling clays. Their distinctive feature is that when heated the thin layers curl up in the form of worms. The interlayer spacing of vermiculite is 14 Å.

Chlorite Group. Common chlorites have 2:1:1 structure and are nonswelling. Because of the 2:1:1 structure they have a rather large interlayer spacing of 14.2 Å even though they are nonswelling.

Minerals of the chlorite group are green because they are commonly derived by the degradation of ferromagnesian minerals rather than feldspar. In addition to the partial replacement of silicon by aluminum in the tetrahedral layer and of aluminum by magnesium in the octahedral layer, Fe, Cr, and Mn may also replace Mg.

Mixed-Layer Clay Minerals. Minerals belonging to this group have inter-stratified layers of different clay minerals. For example, a montmorillonite–halloysite-kaolinite sequence may develop by weathering in tropical moist climates.

Structurally Related Phyllosilicates. Minerals belonging to this group possess a lathlike structure rather than the layered structure common to clay minerals. *Sepiolite* (hydrous magnesium aluminosilicate) and *atapulgite* (hydrous magnesium silicate) are examples of these minerals. They are commonly used in cleaning masonry containing water-soluble salts because of their high cation-absorption ability.

2.2.3 Identification

Clay minerals occur as very fine particles measuring less than 2 μm in diameter. Therefore they do not lend themselves to identification by optical microscopy that is commonly employed to study other minerals. Furthermore, clay minerals do not reveal distinct crystal form; thus their identification by means of electron optics, such as transmission and scanning electron microscopes, is also difficult. X-ray diffractometry (XRD) is the only technique for convenient and accurate identification of clay minerals. However, the clay minerals, as they are often dispersed in carbonate rocks, must be concentrated for X-ray diffraction analysis. A thickness of 0.07 mm to 0.12 mm of clay-mineral mass, often collected upon a glass slide or porcelain plate, is considered optimal. With less thickness the X-ray beam would penetrate the clay-mineral mass and produce reflection from the substrate. If the mount were too thick, then a high background noise would be produced. The purpose of sample preparations is not just to obtain a concentrated mass of the clay minerals, but also to obtain an oriented mount. These features are discussed in the following.

2.2.3.1 Preparation of Sample A sample of rock weighing nearly 50 g is digested in glacial acetic acid. Stronger acids, such as hydrochloric acid, may also be used to dissolve the sample, but there is a likelihood that the structure of the clay mineral may be disrupted in the process. To accelerate digestion in acetic acid air may be periodically drawn out of the flasks using a vacuum pump. To dissolve the entire carbonate portion of the sample, several days are needed during which the acid is replenished periodically after effervescence subsides and the supernatant is carefully removed. The insoluble residue is then washed in distilled water and dried. The dry matter is placed on a 230-mesh screen in order to separate the sand fraction from the silt and clay; the latter pass through the screen.

Water is added to the silt and clay fraction, and the mixture is agitated and then allowed to stay for nearly 6 hours during which most of the silt settles out. The suspended matter is siphoned out and treated with a saturated solution of sodium chloride in which the clay minerals flocculate and are able to settle from the suspension.

As shown previously, interlayer water absorption by clays is an indicator of their structure. However, different cations contain different amounts of water of hydration. To ensure that the interlayer expansion is truly represented in identification procedures, expansion by hydration must be uniform. Therefore, clay mineral slurry is saturated with potassium or magnesium by treating it with a saturated solution of potassium or magnesium chloride and then thoroughly rinsing it with water to remove the chloride solution. This procedure, called *exchange saturation*, makes the slurry a homocationic mass.

Slurry of dispersed clay matter is then mounted on a glass slide or perforated ceramic plate. The latter method is preferred because it allows easy removal of water from the slurry and presumably causes a better orientation of the basal planes of the mineral grains parallel to the substrate. The slurry is dried at room temperature. The thickness of the dry clay film left on the plate can be measured under an optical microscope by the use of a micrometer fitted on the focusing knob. The difference in height between the microscope focus of the surface of the clay deposit and that of the surface of the plate gives the thickness of the clay deposit. The ideal thickness for this example is 0.07 mm to 0.12 mm. The sample mount is now ready for XRD analysis.

2.2.3.2 X-Ray Diffraction Analysis X-ray diffraction analysis determines the crystal structure of minerals, thereby permitting their identification. Minerals are chemical compounds in which the component atoms are arranged in an orderly, repeating, three-dimensional array. In such an array, the repetition of atoms in a straight line can be considered to make rows and rows to make planes. Planes stacked in space produce a space lattice. Each mineral is characterized by a specific space lattice.

The atomic planes in a space lattice can be considered as mirrors. In the optics of visible light, a mirror will totally reflect rays only when it is held at a specific angle to the incident beam; such is also the case for an X-ray beam. However, X rays penetrate deeply into the mineral and encounter many parallel atomic planes separated by a uniform distance, called the *d* spacing. Each plane reflects a portion of the X-ray beam. If the reflected rays are in phase, that is, if *d* is a whole multiple, 1,2,3, . . . *n*, of the wavelength, the intensity of each ray is integrated, producing what is called diffraction (Fig. 2.5). The condition for diffraction is defined by the Bragg equation:

$$n\lambda = 2d \sin \theta \tag{2.4}$$

where θ is the angle of total reflection, λ is the wavelength of X radiation used, and n is the number of planes in the nth order.

The intensity of diffraction is a function of scattering by electrons, the atom, the number of unit cells per unit volume, the volume of the crystal, the temperature, and absorption factors. Therefore, relative diffraction intensities of different sets of planes in a mineral vary.

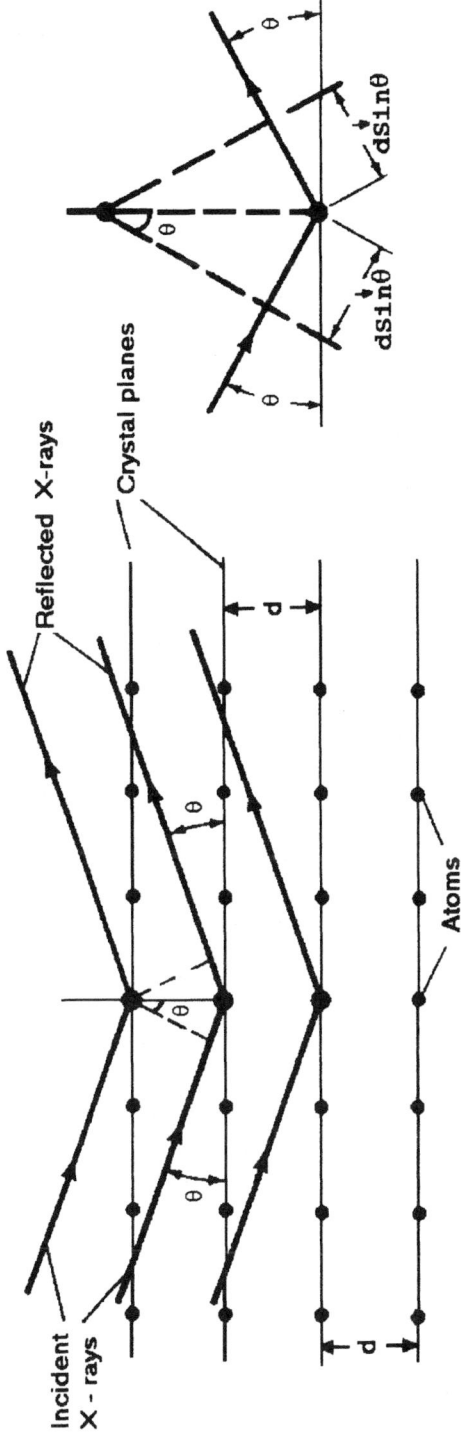

FIGURE 2.5 Geometry of X-ray diffraction. X-ray diffraction is produced when the difference in path length equals 1 or a whole multiple of wavelength. The condition for diffraction is defined by the Bragg equation $n\lambda = 2d \sin \theta$. *Source:* Kane, J. W. and Sternheim, M. M., *Physics*, 2nd ed., New York: Wiley, p. 493. Reprinted by permission of John Wiley & Sons, Inc. Copyright 1983 by John Wiley & Sons, Inc.

Following Eq. (2.4), the X-ray diffraction peaks can be characterized by 2θ or d. The goniometer of the X-ray diffractometer that rotates the sample through various angles of incidence of the X-ray beam is calibrated to read the 2θ values. Obviously, when 2θ is known, d can be easily calculated from the Bragg equation. Table 2.2 gives 2θ degrees and corresponding d in the range of 5 to 45 2θ for the commonly used radiation of wave length 1.5148 λ. In Appendix C, we give a program that enables one to find 2θ or d in any range and for any X-ray radiation used.

TABLE 2.2   Conversion of d and 2θ Value for X-radiation Wavelength (λ) 1.5418 Å

2θ	d (Å)	2θ	d (Å)
5.0	17.6733	25.0	3.5617
5.5	16.0677	25.5	3.4930
6.0	14.7298	26.0	3.4270
6.5	13.5978	26.5	3.3634
7.0	12.6277	27.0	3.3023
7.5	11.7869	27.5	3.2434
8.0	11.0513	28.0	3.1866
8.5	10.4023	28.5	3.1318
9.0	9.8255	29.0	3.0789
9.5	9.3095	29.5	3.0279
10.0	8.8451	30.0	2.9785
10.5	8.4250	30.5	2.9308
11.0	8.0431	31.0	2.8847
11.5	7.6945	31.5	2.8400
12.0	7.3750	32.0	2.7968
12.5	7.0811	32.5	2.7549
13.0	6.8099	33.0	2.7143
13.5	6.5588	33.5	2.6749
14.0	6.3256	34.0	2.6367
14.5	6.1086	34.5	2.5996
15.0	5.9061	35.0	2.5636
15.5	5.7167	35.5	2.5287
16.0	5.5391	36.0	2.4947
16.5	5.3724	36.5	2.4617
17.0	5.2155	37.0	2.4295
17.5	5.0676	37.5	2.3983
18.0	4.9279	38.0	2.3679
18.5	4.7959	38.5	2.3383
19.0	4.6708	39.0	2.3094
19.5	4.5521	39.5	2.2813
20.0	4.4394	40.0	2.2540
20.5	4.3323	40.5	2.2273
21.0	4.2302	41.0	2.2013
21.5	4.1330	41.5	2.1759
22.0	4.0402	42.0	2.1511
22.5	3.9515	42.5	2.1270
23.0	3.8667	43.0	2.1034
23.5	3.7856	43.5	2.0804
24.0	3.7078	44.0	2.0579
24.5	3.6333	44.5	2.0359
		45.0	2.0145

The instrument for X-ray diffraction is called a diffractometer (Fig. 2.6), which consists of a high-voltage power source that excites X-rays from (1); a target (2), often a copper filament enclosed in an evacuated X-ray tube; a slit system (3) to focus X rays onto the mineral sample, which is held by an arm of goniometer (4), a device that can rotate so that different planes of the crystal lattice can be oriented optimally with respect to the X-ray beam; and a detector (5) mounted on the goniometer, which gives digital values of intensities and passes these on to an integrated circuit control (6), which can read counts per time interval and record these values in graphic form, producing a diffractogram. Figure 2.7 shows a Philip diffractometer.

The procedure for X-ray diffraction analysis can be summarized as follows:

1. *X-Ray Diffraction pattern.* XRD pattern, or diffractogram, of an oriented mount of dried clay slurry for $2\theta = 5°$ to $25°$ is made using the wavelength of Cu $K\alpha$ radiation ($\lambda = 1.5418$ Å). The peaks are identified by 2θ and d. These peaks are then compared with known standards given in the American Society for Testing and Materials (ASTM) powder diffraction files. Table 2.3 gives the three intense XRD reflections of the principal clay-mineral species. It is apparent from this table that the minerals smectite, vermiculite, and chlorite have maximum intensity peaks at nearly $2\theta = 14°$. If any of these minerals is present, then further identification is necessary to pinpoint the mineral for which the following treatments are needed.

2. *"Glycolation".* Glycolation is the exposure of the sample to ethylene glycol vapor. In this procedure the sample is placed in a desiccator containing ethylene glycol, which is then heated to nearly 60°C. Alternatively, a few drops of ethylene glycol are added to the dry sample. The relatively larger ethylene glycol molecules replace the interlayer water molecules, if present, causing the lattice to expand. Another diffractogram is then taken and the maximum intensity peak measured. Chlorite, being nonswelling, will show no shift in the peak position. If the shift does occur, then the smectite and vermiculite can be distinguished by the following procedure.

3. *"Glyceration" or treatment with glycerol.* Drops of glycerol solution in water in a ratio of nearly 10:1 are placed upon the sample. Glycerol is absorbed by smectite but not by vermiculite. Thus if a change in the maximum intensity peak position occurs, the presence of smectite is indicated.

4. *Heating to 180°C.* In general, swelling and nonswelling clays can be distinguished by heating to 180°C, the temperature at which they lose bonded water. If the water is lost, d will be reduced, which can be seen from diffractograms taken before and after the heat treatment.

5. *Heating to over 500°C.* To firm up identification the sample is heated to temperatures above 500°C at which certain clays become amorphous, that is, they lose crystal structure. Kaolin, for example, becomes amorphous at 550°C.

Figure 2.8 shows the X-ray diffraction traces of common clay minerals as identified by the techniques just listed.

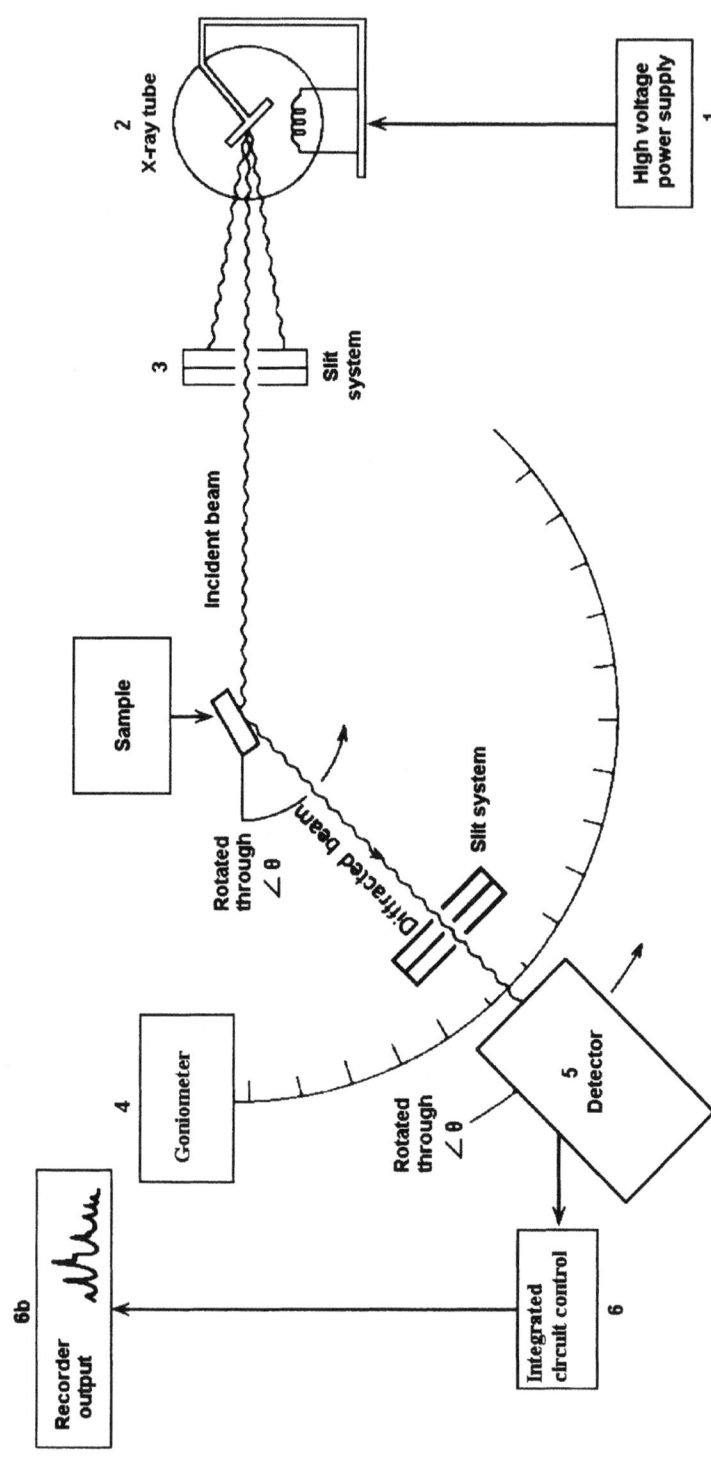

FIGURE 2.6 Schematic of an X-ray diffractometer system. *Source:* Modified from Liebhafsky, H. A., and Pfeiffer, H. G., "X-rays and electron analytical chemistry with emphasis on instrumentation," in *Chemical Instrumentation-II*, G. W. Ewing, ed., Washington, D.C.: American Chemical Society, 1977, p. 55.

a

b

FIGURE 2.7 Philip X-ray diffractometer: (a) Power supply and goniometer (right) and an integrated control panel (left) and (b) X-ray tube, camera with a sample in position, and goniometer with curved-crystal monochrometer.

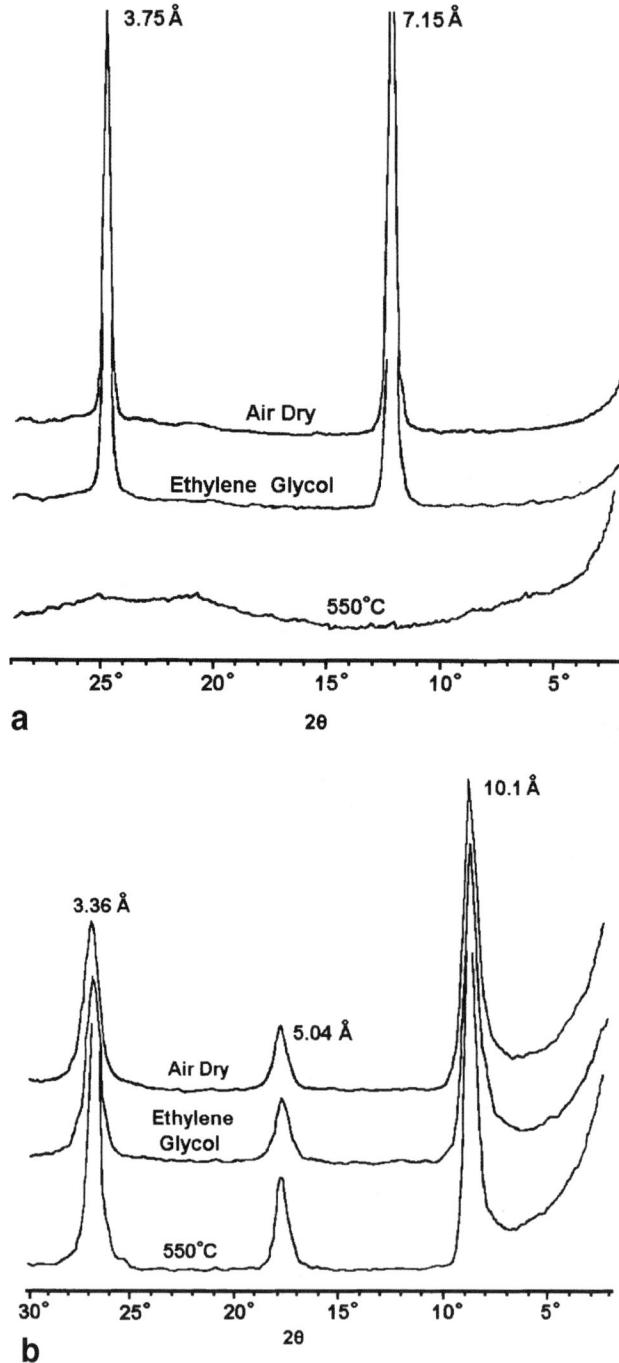

FIGURE 2.8   Identification of clay minerals by various treatments. Maximum intensity peaks are plotted; numbers on peaks represent d, corresponding to 2θ values shown along the scale.

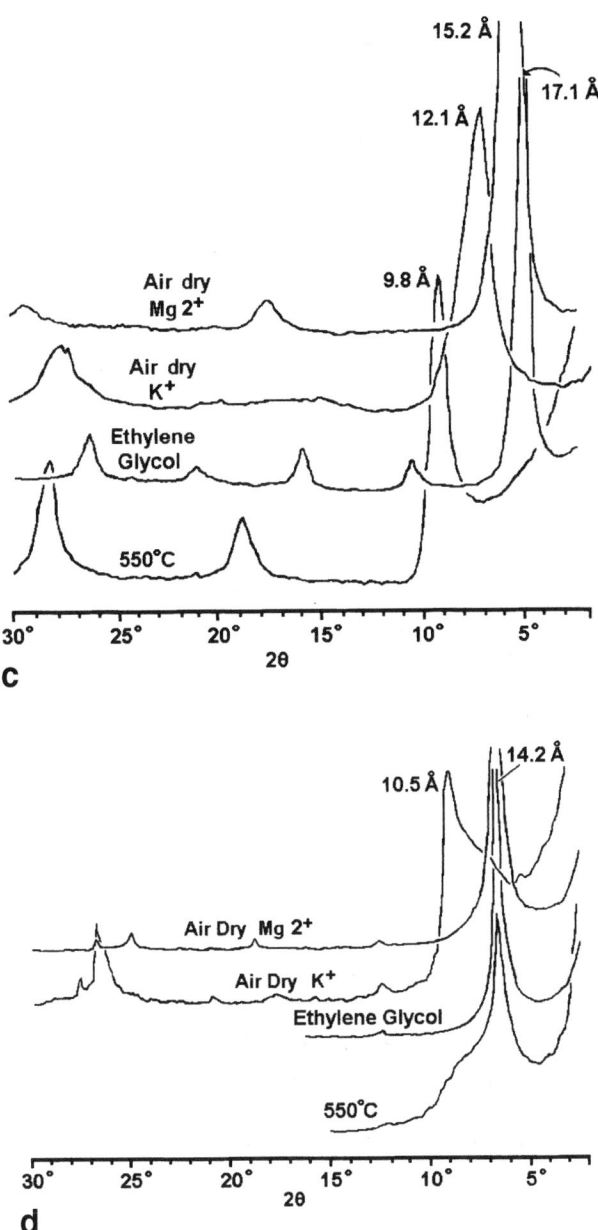

FIGURE 2.8 (Continued). (a) Kaolin, (b) illite, (c) montmorillonite, (d) vermiculite, and (e) chlorite. *Source*: Eslinger, E. and Deaver, D., *Clay Minerals*, Society of Economic Paleontologists and Mineralogists, 1988.

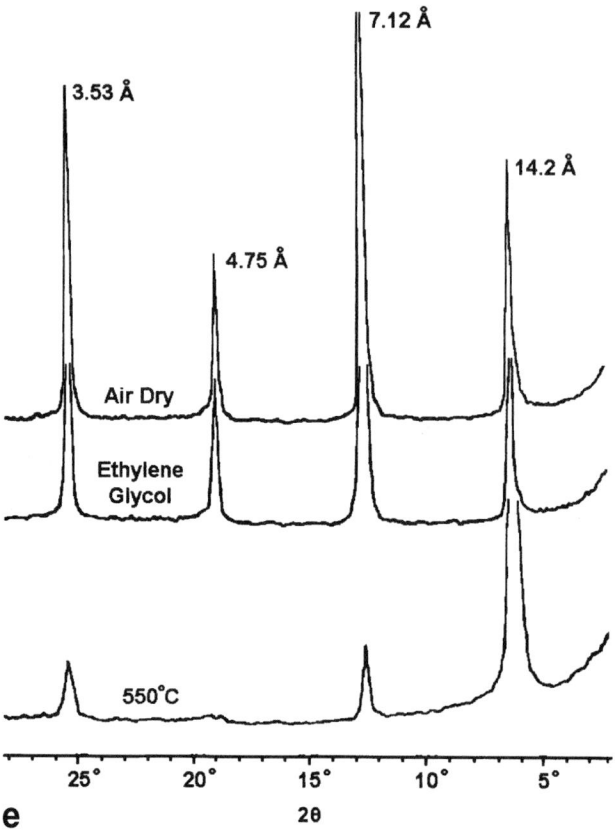

FIGURE 2.8. (Continued)

2.3 WATER-SOLUBLE MINERALS

Water-soluble minerals are those that are present in the water in ionic form before crystallization. They are also called evaporites or evaporite minerals due to their crystallization from water by evaporation. Common ions in seawater are Na^+, K^+, Ca^{2+}, Mg^{2+}, Cl^-, SO_4^{2-}, etc., and common evaporite minerals are halite (NaCl), gypsum ($CaSO_4 \cdot 2H_2O$), etc. According to this definition calcite ($CaCO_3$), the major mineral of carbonate rocks, is also an evaporite mineral; however, due to its crystallization at normal seawater temperature it is not usually included in the evaporite minerals.

The sequence of crystallization of evaporite minerals is determined by their solubility in water (Chap. 4, Table 4.1). For example, calcium carbonate has the lowest solubility (hence it is commonly considered as a non-water-soluble mineral) and it is the first to crystallize, followed by gypsum and halite. The temperature gap between the crystallization of these minerals is large; this is why they cannot genetically

TABLE 2.3 X-Ray Diffraction Peaks for the Three Most Intense XRD Reflections for the Principal Clay-Mineral Species

Clay Mineral	d (Å)	2θ (deg)	Intensity
Kaolinite	7.15^a	12.4	100
$Al_4(Si_4O_{10})(OH)_8$	3.58	24.9	100
	2.30	39.2	60
Illite	10.08^a	8.8	100
$K, Al_2(Al, Si_3O_{10})(OH)_2$	4.9	17.7	90
	3.3	26.7	90
Smectite	12.0^a–15	7.3–6.9	100
$(Ca,Na)_{0.35-0.7}(Al,Mg)_2(Si, Al)_4O_{10}(OH)_2 \cdot nH_2O$	6.0–7.5	14.7–11.8	60
	4.0–5.0	22.2–17.8	40
Vermiculite	14.36^a	6.15	100
$Mg_{11}Al_5FeSi_{11}O_{42} \cdot 40\ H_2O$	7.2	12.3	60
	4.8	18.5	60
Chlorite	14.4^a	6.14	100
$(Al,Mg,Fe)_3(Al,Si_4O_{10})(OH)_2 (Mg, Fe)3\cdot(OH)_6$	7.15	12.38	60
	4.79	18.52	40

^{a}Basal plane spacing.

coexist in a rock. Figure 2.9 shows the separate sites at which carbonate and sulfate (gypsum) build up would occur. However, because these sites can be geographically quite close to each other, gypsum can easily wash into the environment of carbonate deposition, resulting in its incorporation in the carbonate sediment. By the same process, halite and other water-soluble salts, if forming in close proximity, can occur in carbonate rocks. This explanation also clarifies the statement made in the Introduction of this chapter, "the low solubility of calcium carbonate precludes simultaneous precipitation of other evaporite minerals, which are more soluble."

Carbonate Basin   Tidal Flat

FIGURE 2.9 Noncarbonate minerals in carbonate sediments. The carbonate sediments form from normal seawater. However, evaporite minerals, such as gypsum, that crystallize nearby from hypersaline water may be washed into the carbonate sediment. A carbonate basin and a nearby tidal flat where the crystallization of gypsum (and possibly halite) may occur are shown.

TABLE 2.4   X-Ray Diffraction Peaks for the Three Most Intense XRD Reflections for Calcite, Gypsum, and Halite

Clay Mineral	d (Å)	2θ (deg)	Intensity
Calcite	3.04	29.4	100
$CaCO_3$	2.29	39.3	18
	2.10	43.0	18
Gypsum	7.56	11.7	100
$CaSO_4 \cdot 2H_2O$	3.06	29.1	57
	4.27	20.8	51
Calcium sulfate	2.98	29.9	100
subhydrate			
$2CaSO_4 \cdot H_2O$	2.78	32.2	100
	5.98	14.8	90
Anhydrite	3.50	25.4	100
$CaSO_4$	2.85	31.4	33
	2.33	38.6	22
Halite	2.82	31.9	100
NaCl	1.99	45.5	55
	1.63	56.4	15

Water-soluble minerals are commonly present in a very small quantity in limestone and when present are uniformly distributed in the mass of the rock. Therefore, they must be concentrated in a larger mass to allow their identification.

The water-soluble minerals can be extracted from the rock mass by placing a preferably crushed sample of rock in deionized water. Drops of this solution are then placed on a glass slide, which is then heated to a low temperature of 30°C to 40°C so that crystals of the mineral can form. After the water has evaporated the glass slide can be viewed under an optical or an electron microscope to identify the minerals and analyzed by X-ray diffraction for confirmation. Table 2.4 gives the X-ray diffraction peaks for the three most intense diffractions of the principal evaporite mineral species found in carbonate rocks. In practice, the diffraction patterns may show slightly shifted peaks and intensities, or even the order of intensity, that differ from values shown in the table. This is due to inconsistencies of instrumental settings and orientation of the crystals on mounts as they crystallize from the solution. In Table 2.4 we also show the minerals anhydrite and calcium sulfate subhydrate, which though absent in carbonate rocks may be present in the dried leachate formed by conversion of gypsum.

SUMMARY

Clay and evaporite minerals are occluded in the carbonate sediments by having been transported from areas outside the carbonate basin. Clay minerals are the end products of weathering of igneous rocks. Evaporite minerals crystallize from marine waters that become saturated with the ions that make up these minerals. The identification of both evaporite and clay minerals is essential for understanding the weathering patterns of carbonate rocks. X-ray diffraction analysis is described, which is the method of choice for identification of these minerals.

CHAPTER 3

Structural Deformation of Carbonate Rocks

3.1 INTRODUCTION

This chapter deals with the natural processes of structural deformation of rock and with the change in engineering properties brought about in experiments simulating outdoor weathering.

Carbonate rocks are originally deposited as horizontal layers of soft sediment at the bottom of the sea and hardened over time. But on continents these layers often appear tilted and fractured. The fractures, called joints, are one of the many manifestations of structural deformation caused by mechanical stresses associated with the uplifting of rocks above sea level and the behavior of these rocks as brittle materials. The study of joints is particularly useful in the case of monuments, which are large and carved in the country rock, for example, the Sphinx. We will show how to represent joints in a stereographic projection so that their spatial orientation and thus their influence upon weathering can be recognized. Another aspect of fractures is their creation by salt crystallization in rock pores. We will simulate the natural process of salt crystallization and determine the change in the elastic modulus, which is a measure of the rock's resistance to weathering. Finally, if the rock is sheltered from rain in outdoor exposure, weathering produces a crust upon the surface. We will describe a method by which the change in the compressive strength of the zone of weathering can be measured so that this property can be used to evaluate the magnitude of weathering and as a performance criterion if the rock is treated for improved durability.

3.2 CARBONATE ROCKS AS BRITTLE MATERIAL

Solid substances behave as brittle or ductile materials when stress is applied to them. Figure 3.1 shows a universal testing machine in which a sample can be subjected to stress.

Stress is defined as force (F) per unit cross-sectional area (A)

$$\sigma = \frac{F}{A} \tag{3.1}$$

49

FIGURE 3.1 Universal testing machine: An instrument used for measurement of compressive and tensional strength of materials. In this machine the part on the left consists of a stage on which the sample to be tested is placed; its upper portion is a hydraulic press that applies pressure on the sample. The pressure is shown by the gauges on the right.

and often expressed as pounds per square inch (psi), kilograms per square centimeter (kg/cm^2) or newtons per square meter (N/m^2). To compare studies expressing stress in different units, the conversion factors are 1 kg/cm^2 = 9.8×10^4 N/m^2 = 0.0703 psi.

Three kinds of stress are commonly known: compressive stress, in which forces acting towards each other squeeze a body; tension stress, in which forces acting away from each other tend to stretch a body; and shear stress, in which forces push one side of a body past the adjacent part, also tending to stretch a body. When subjected to any of these stresses, materials may undergo deformation.

The change in length of a cylinder in compression or tension stress is proportional to its length. For example, a cylinder of length l subjected to a compressional force F shortens a distance Δl, then strain ε is defined as the fractional change in length

$$\varepsilon = \frac{\Delta l}{l} \tag{3.2}$$

which is dimensionless.

The relation between the stress and strain for some materials can be found experimentally. For example, a sample being squeezed in a compression test will experience a decrease in length. If the force is released, the sample may regain its length. This strain is termed *elastic strain*. If the force is increased some material will finally break (Fig. 3.2). Such materials are termed brittle materials. The force at which

FIGURE 3.2 Behavior of marble in unconfined compression. Showing fractures in a marble resulting from uniaxial compressional stress, that is, the sample is unsupported on the sides as the load is applied vertically.

the sample will break is called its compressive strength. However, certain materials, after the limit of elastic strain is reached, will not break but will flow. These are ductile materials. The mechanical behavior of a material can be shown by stress–strain diagrams (Fig. 3.3). Carbonate rocks behave as brittle materials when formed near the

FIGURE 3.3 Behavior of materials under stress. When compressional force is applied, materials initially shorten (strain). In elastic deformation, rocks regain their original length if the stress is removed. The increase of stress beyond the elastic limit can result in plastic flow before breakage in *ductile* materials. *Brittle* materials spontaneously rupture after the stress exceeds the elastic limit. Rocks act as brittle materials when present near the earth's surface and ductile when buried deep inside the earth.

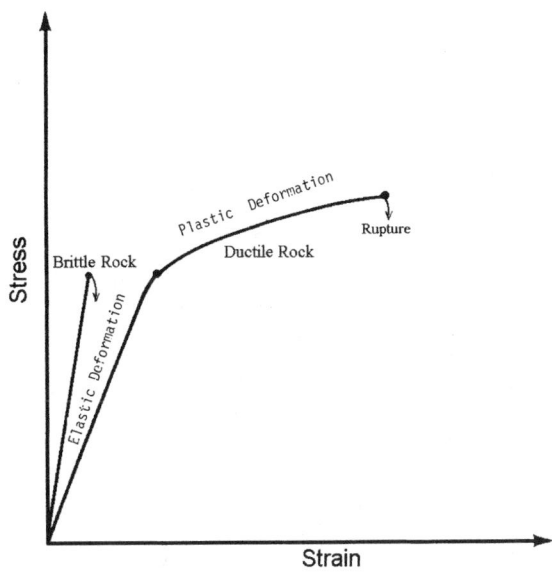

earth's surface (limestone, dolomite), but undergo plastic deformation when subjected to high temperature and pressure at great depth in earth's interior (marble).

Tensile strength represents the force at which material will rupture when subjected to tensional forces. The tensile strength of carbonate rocks is generally less than one-fourth their compressive strength.

3.2.1 Joints

Joints are fracture planes in a rock. In a mass of rock, the overlying burden subjects the deep-seated portion to gravitational pressure, termed lithostatic pressure. This force is operating equally from all directions at any point in the rock mass and is termed as confined pressure. When the rock is lifted upward by forces within the earth, erosion occurs or, similarly, in mining when upper layers are removed the lower layers may expand and become fractured parallel to the surface. These fractures are called exfoliation joints. Other joints, called normal joints, may also appear which are a result of the differential removal of the confining pressure. They may be gently sloping to nearly vertical.

However, in natural uplift of limestone and dolostone, exfoliation joints are uncommon because the overlying burden is thin and the unloading by erosion is slow. The mining operations, however, must be carried out slowly so that the rock is slowly destressed. The following relates mainly to the normal joints which are the most common structural feature of the carbonate rocks as seen in natural outcrops (Fig. 3.4).

Joints, as any planar structure on earth can be described by attitude, which consists of strike and dip. Strike is the azimuth, that is, angle from north and dip is the angle of inclination with respect to the horizontal surface. The concept of attitude can be easily understood by reference to a podium (Fig. 3.5) where the ridge represents strike and slope the dip. Since the ridge runs parallel to EW direction in this figure, the strike is 90° and the dip 45° towards south. The attitude of this surface then can be written as 90/45S. Attitude is measured by the use of a geologic compass (Fig. 3.6) which consists of a freely suspended magnetic needle to measure strike and a clinometer to measure dip.

FIGURE 3.4 Joints in limestone at the Giza Plateau, Egypt. These joints formed as the strata were uplifted in Eocene tectonism. Note that the joints are present in the hard (competent) layers and appear to be absent from the interlayered softer layers.

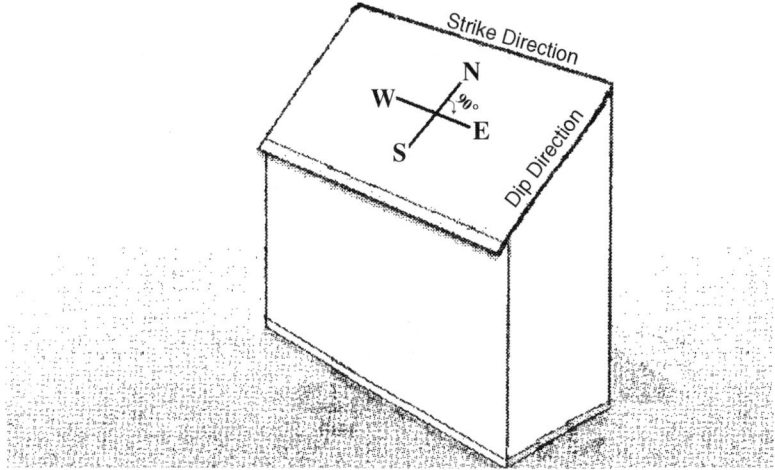

FIGURE 3.5 Attitude of planar surfaces. Planar surfaces, such as joints, can be described by their attitude, which consists of strike and dip measurements illustrated here by comparison with a podium. Strike is given by the orientation of horizontal direction with respect to the geographic NS direction, shown by the ridge of the podium. The dip is measured at a right angle to the strike and given as the angle of tilt from horizontal.

Joints often criss-cross rock layers. They accelerate rock decay as the agents of weathering can easily penetrate the rock mass. Their study is particularly useful for large monuments carved out of the native rock such as the Great Sphinx. Figure 3.4 represents an area outside the Sphinx but the geological strata are the same as those that appear at the Sphinx.

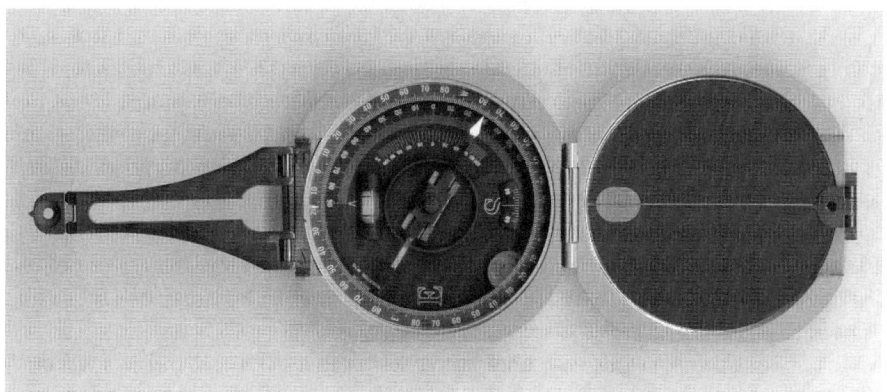

FIGURE 3.6 Geological compass: An instrument used to measure strike and dip. To measure strike of a planar surface, the compass is held lengthwise such that it is horizontal. To measure dip, the compass is rotated so that the long direction is now held against the slope of the plane.

3.2.2 Stereographic Projections to Represent Joints

In a region joints follow definite spatial patterns which can be recognized only by measuring as many joints as possible in the field and representing these in some graphical form. Stereographic projections are devices commonly used to represent joints.

Projections are diagrams, which represent earth features on a plane surface. They are produced by passing lines from various points on the earth's surface to their intersection with a plane parallel to earth's surface. A stereographic projection, or stereonet, is often produced by projecting the points of intersection of meridians and parallels on the southern hemisphere upon the equatorial plane (Fig. 3.7). Thus, the outer circle of the stereonet represents the equator and the center the south pole. These directions have no further meaning in the use of stereonet to depict the attitude of surfaces. However, to show a surface such as a joint the topmost point in the projection is considered the north direction.

To understand the application of stereonet to represent joints it is necessary that one visualizes joints as planes extending into the southern hemisphere of a globe. For example, consider the plane *ABCD* with a NE strike and SE dip [Fig. 3.8(a)]. The intersection of this plane with the surface of the sphere is its spherical projection shown as ABE. This spherical projection, when projected on the equatorial plane is the stereographic projection *AOBF* [Fig. 3.8(b)].

Now, let us go back to a stereonet such as one shown in Fig. 3.7. In order to represent joints, a stereonet is pasted on a hard cardboard surface and a pin, such as a thumbtack, projects from bottom outwards through the center of the net. Tracing paper is then placed upon the net on which the outer circle of the underlying net is drawn and the N direction marked. Now, if we want to plot a joint 25°/20SE, we rotate the tracing paper towards west (anticlockwise) by 25° as marked on the net. Then we draw an arc from N on the tracing paper passing through the E–W line at 20° inwards from outer circle. This arc is the cyclographic trace of the required joint (Fig. 3.9).

When a large set of joints must be represented, many cyclographic traces are cumbersome to draw. However, the plot of poles to the cyclographic traces provides a convenient means to represent joints. Poles can be plotted by obtaining the position of the cyclographic trace, but not drawing it, and marking a point on the east west line 90° from the trace (Fig. 3.9). Thus, the joints are represented by poles in a stereographic net, often the Wulff net shown in Figure 3.10.

A diagram of poles of joints thus produced is called a Pi-diagram. From a Pi-diagram then a general sense of attitude of the joints in an area can be ascertained. To obtain numerical density distribution of joints, however, an equal area projection is needed.

In a stereographic net the directions as they occur in the field are truly represented. An equal area projection, however, is constructed in such a way that areas on the map are equal to the corresponding area in the field. Figure 3.11 shows a Lambert meridional equal-area projection.

The distribution density of poles on an equal area projection can be shown by means of contours. To draw contours, a *point counter* is used, which is essentially a

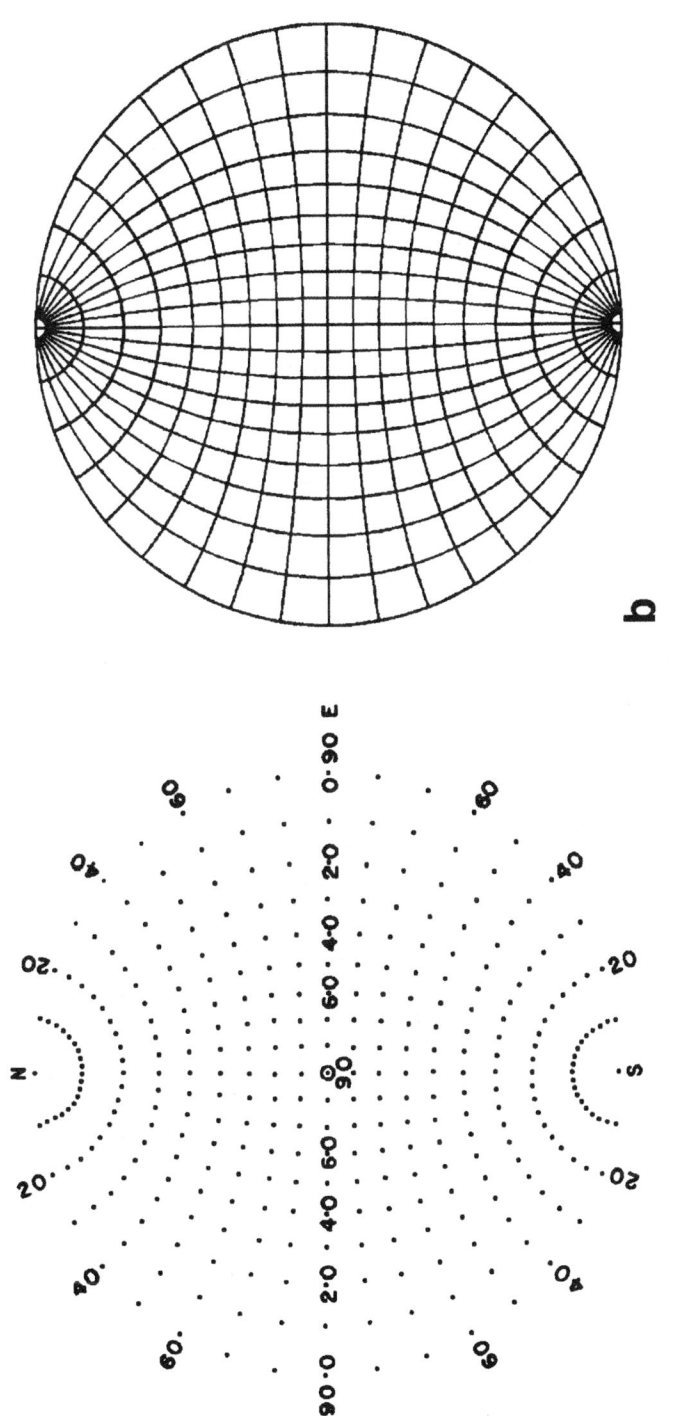

FIGURE 3.7 Development of a stereographic net. A stereographic net is produced by projecting meridians and parallels of the southern hemisphere onto the equatorial plane. Notice that geographical direction is not marked on the stereonet. For representing joints in a stereonet, tracing paper is placed on the net and the N direction marked on it coinciding with the top of the net.

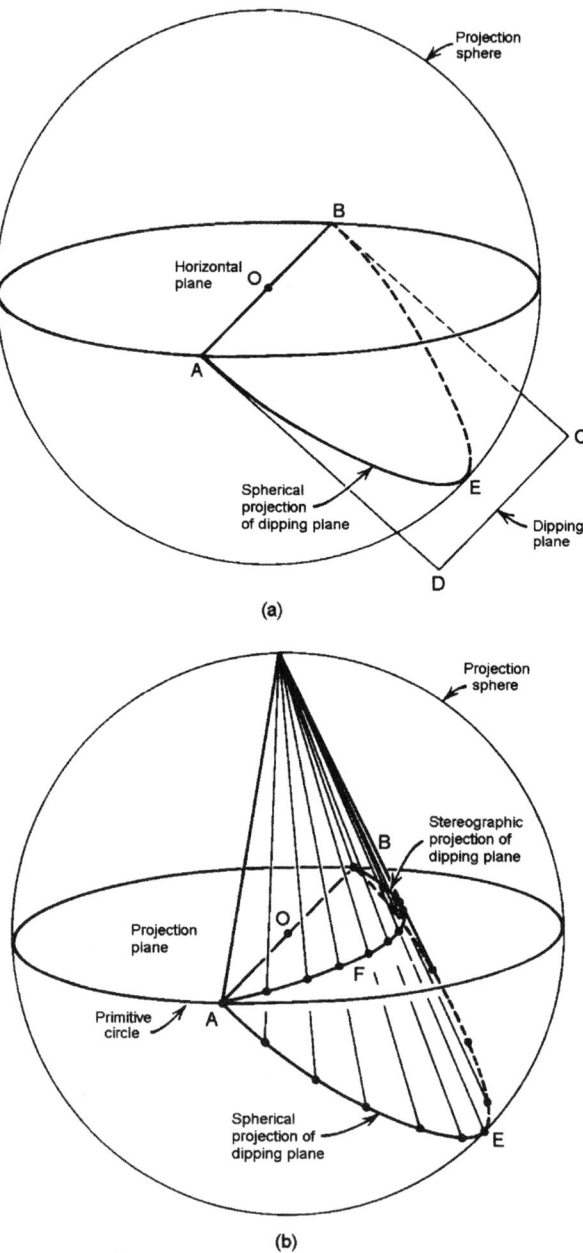

FIGURE 3.8 Representation of joints, or in stereographic projection. The joint is shown as the dipping plane *ABCD* along with (a) its spherical projection and (b) the stereographic projection *AOBF*. *Source*: Hobbs, B. E., Means, W. D., and Williams, P. F., *An Outline of Structural Geology.* New York: Wiley, p. 484. Copyright 1976. Reprinted by permission of John Wiley & Sons.

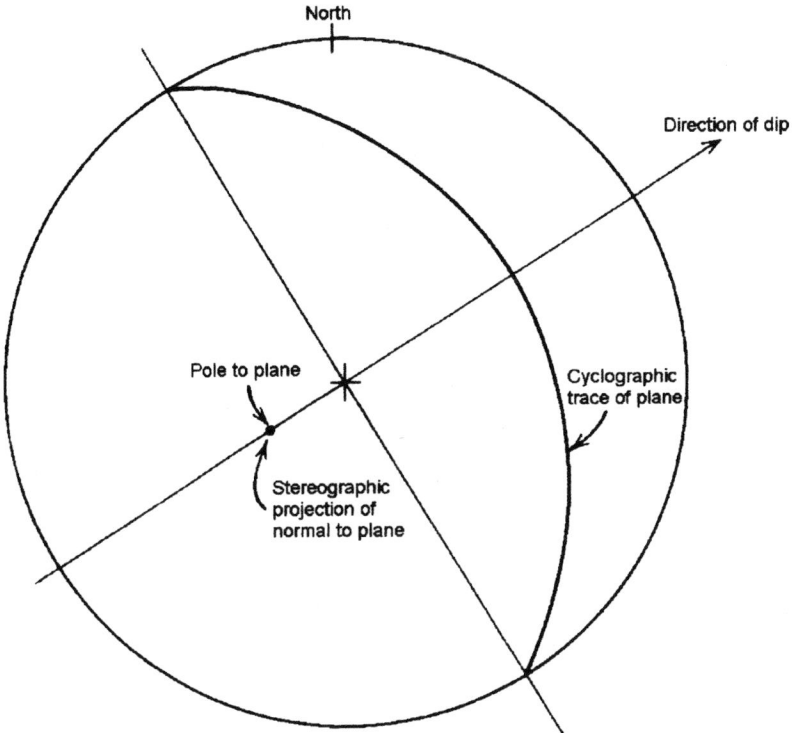

FIGURE 3.9 Cyclographic trace and pole to a plane. The cyclographic trace is equivalent to curve *BFA* in Figure 8(b). *Source:* Hobbs, B. E., Means, W. D., and Williams, P. F., *An Outline of Structural Geology.* New York: Wiley, p. 491. Copyright 1976. Reprinted by permission of John Wiley & Sons.

hollow circle in some stiff object, such as a Plexiglas, such that the area of the circle is 1 percent that of the area of the net [Fig. 3.12(a)]. For instance, if the radius of the net is 10 cm then the radius of the counter is 1 cm. A 1-cm grid system is placed on the net on which the poles have been plotted. The counter is then moved one centimeter from left to right and up and down until the entire surface of the net is covered. The number of poles in a circle at each stop is recorded at the center of the circle. For points near the circumference of the net a peripheral counter is used which has two circles one on each end. The sum of points covered by the antipodal circles is recorded in both circles. The points thus recorded within each circle are converted to percentage figures by their relationship to all points on the net. For example, total number of joints plotted is 100 and 4 joints are present in the counter then 4 percent joints have the orientation as shown by their poles. Contours are drawn which join points of same value. Figure 3.12(b) shows such contours indicating that these are low-angle joints trending generally NEE and dipping towards NNW. Their general attitude can now be given as 70/20NW.

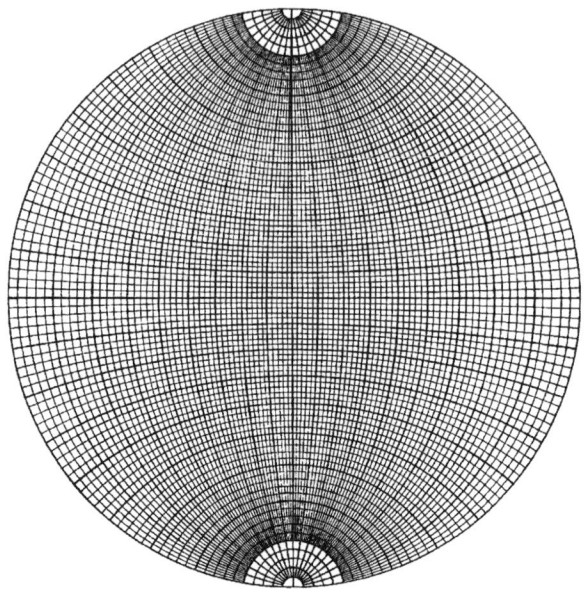

FIGURE 3.10 Wulff net. The Wulff net is a stereographic projection of the great circles of longitudes and latitudes at 2° intervals, the projection points lying on the equatorial plane of the sphere. The Wulff net is an equiangular projection in which the angular relationships are truly maintained.

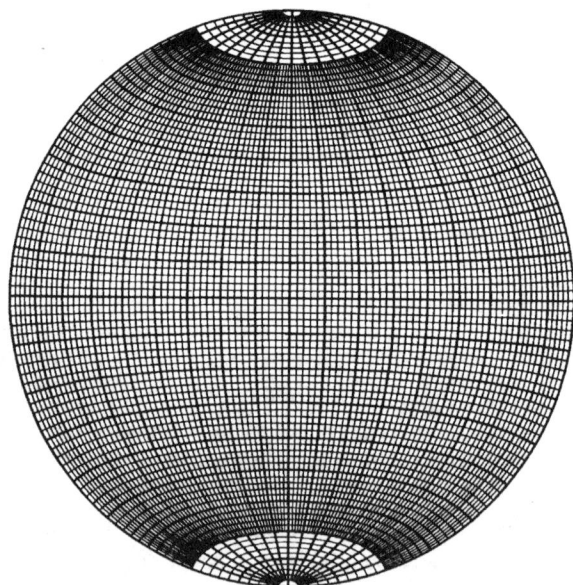

FIGURE 3.11 Lambert meridional equal-area projection. In this projection the directions become distorted, but areas are truly represented.

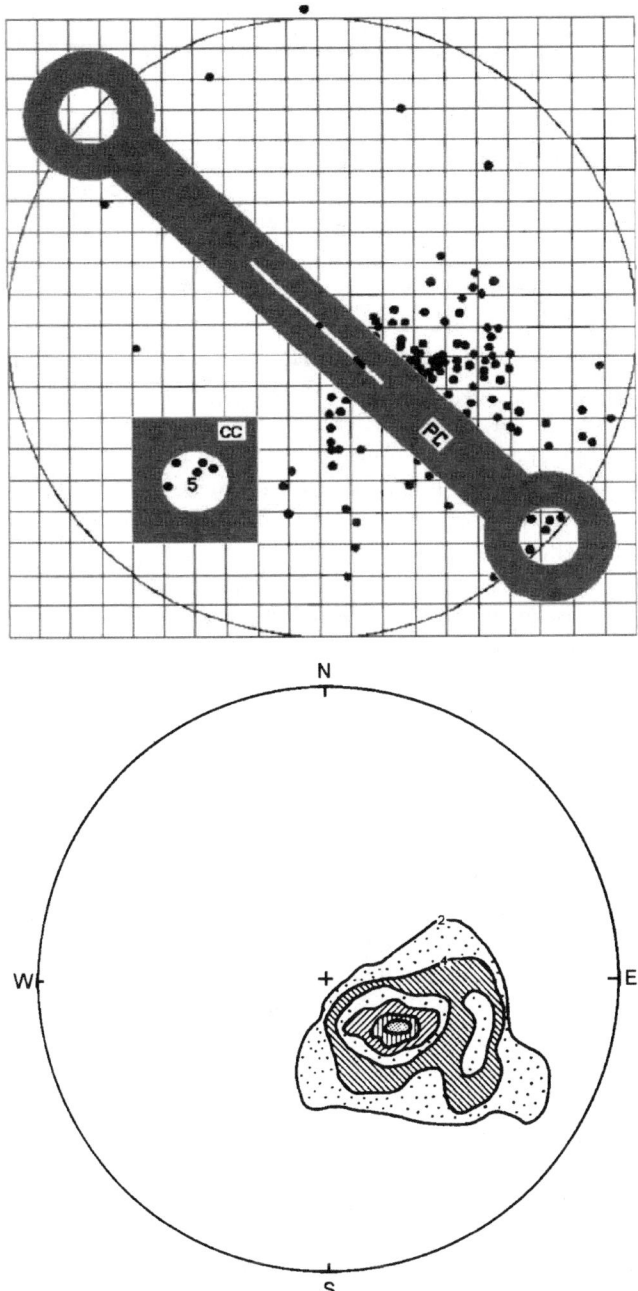

FIGURE 3.12 Measuring density of poles to joints. (a) Point (CC) and peripheral counter (PC) measuring density distribution of poles and (b) contour map of pole density. (b) indicates that most joints have a nearly NE–SW strike and 20° NW dip. The contours are 2%, 4%, 6%, 8%, 10%, and 12%.

In Chapter 9 we show joint patterns and their influence on the durability of the Sphinx. Such studies are useful in determining causes of deterioration but also guide the selection of sites for layout of monolithic structures.

3.3 DEFORMATION BY SALT CRYSTALLIZATION

We describe in Chapter 4 that weathering can be mechanical or chemical whereby the structural properties are altered in the zone of weathering which is often very thin. Therefore, structural properties of the zone of weathering, such as compressive and tensile strength, which are measured on bulk materials commonly by engineering techniques, cannot be directly measured. However, methods can be developed to correlate the properties of the weathered and sound stone. In the following we will show as to how the deformation induced by salt crystallization (mechanical weathering) and by natural crust formation (chemical weathering) can be expressed by the engineering properties. It is clear that this knowledge can be used both for the selection of sound materials as well as for measuring the magnitude of change caused by the natural weathering.

It is known that the inherent presence of water-soluble salts in stone promotes disintegration. This knowledge has been used in devising methods to predict the durability of porous stone. The German Standard (DIN 52111) and the American Society of Testing and Materials (ASTM C 88) use the crystallization of sodium sulfate for this purpose. While the mechanics of damage by salt crystallization is discussed in Chapter 4, here we will describe experiments, following DIN and ASTM methods, in which deformation was induced by depositing sodium sulfate (Na_2SO_4) in samples of Indiana limestone and the changes in strain, compressive strength, tensile strength, and Young's modulus measured with increasing cycles of salt deposition.

As is common in engineering practice for such testing, samples were used with the dimensions 5×5 cm on the surface and 7.5 cm long, and cores nearly 7.5 cm long and 3.75 cm in diameter. They were immersed in saturated solution of sodium sulfate at 20°C (prepared by using nearly 215 g of anhydrous sodium sulfate per liter of water) for a period of 24 h. The specimens were then dried at 105°C and immersed again in sodium sulfate solution. Thus, each cycle of wetting and drying increased the salt content in the pores thereby producing increasingly higher pressure in the subsequent cycle of wetting. Figure 3.13 shows deterioration of samples in 1–5 cycles of sodium sulfate crystallization.

The samples expand as salt crystallization occurs in the pores. To measure change in length (strain), a 1-cm-rod was screwed in at the bottom of a square specimen that projected out at the bottom of the beaker in which the specimen was immersed. This rod was threaded into a steel plate. Another rod was placed at the top of the specimen, which barely touched a micrometer, or a displacement transducer. The setup is shown in Figure 3.14. The specimen was coated with epoxy from all direction except the top so that the solution entered the pores from top only. After sodium sulfate had been deposited in pores in the first immersion, the second and subsequent immersions

FIGURE 3.13 Disintegration of rock by sodium sulfate crystallization. The samples are Indiana limestone, also called Bedford limestone. The numbers on the samples indicate the number of cycles of crystallization. *Source*: M. Sharifinassab, An experimental investigation of concrete and limestone deterioration by crystallization of sodium sulfate, M.S. thesis, University of Louisville, 1981, p. 57.

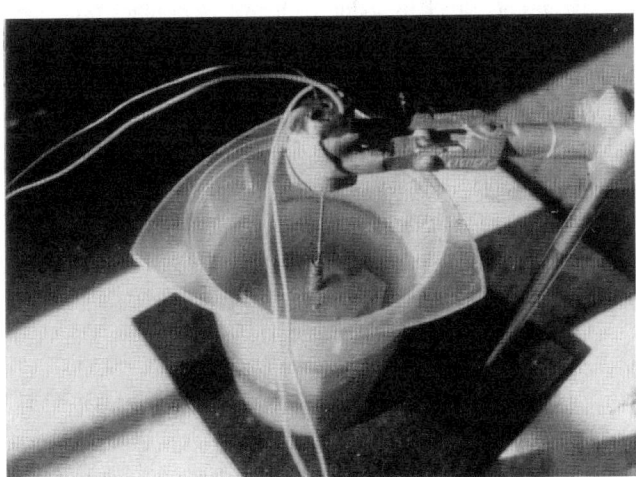

FIGURE 3.14 Expansion of rock by sodium sulfate crystallization. The experimental setup to measure expansion is shown. Note that the bottom of the sample was anchored to a steel plate by a rod while another rod projected above the sample surface and touched the displacement transducer. The sample was coated with epoxy on all sides except the top so that the depth of salt impregnation could be measured. *Source*: M. Sharifinassab, An experimental investigation of concrete and limestone deterioration by crystallization of sodium sulfate, M.S. thesis, University of Louisville, Fig. 5, 1981.

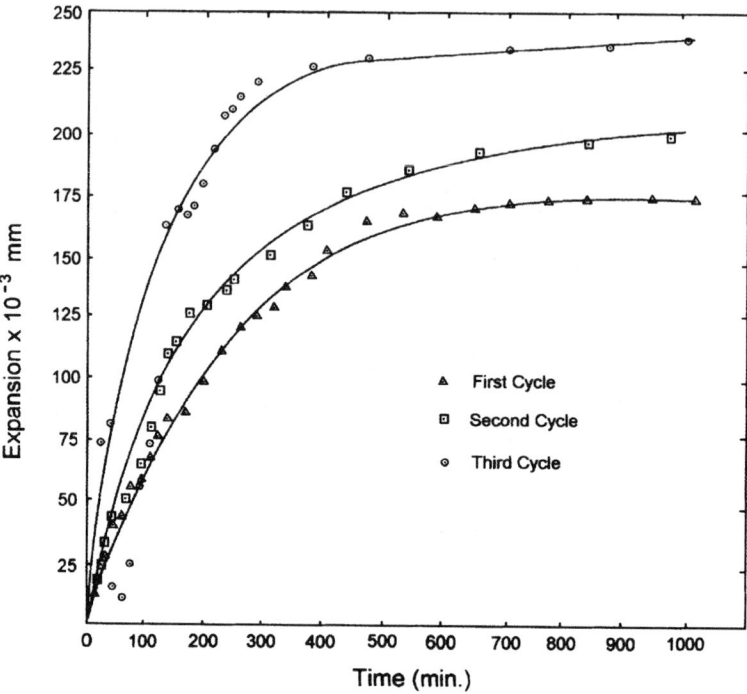

FIGURE 3.15   Expansion of rock by sodium sulfate crystallization. The linear expansion is measured as a function of time. *Source:* Cassaro, M. A., Gauri, K. L., Sharifinassab, M. and Sharifian, A. in K. L. Gauri and J. A. Gwinn (eds.), *On the Strength and Deformation Properties of Indiana Limestone and Concrete in the Presence of Salts*, The University of Louisville, 1982, pp. 57–76.

showed an increase in length giving change in length with cycles of crystallization (Fig. 3.15).

3.3.1   Change in Compressive Strength and Tensile Strength

Treated samples that had gone through one, two, and three cycles of crystallization and virgin core samples were then subjected to standard engineering tests for uniaxial compressive strength and tensile strength in a universal testing machine. Compressive strength was measured by increasing the load on a vertically held sample until the sample failed (Fig. 3.2). The tensile strength was measured indirectly by the split cylinder tensile test (Fig. 3.16) and expressed by

$$\sigma_T = \frac{2P}{\pi DL} \tag{3.3}$$

FIGURE 3.16   Tensile strength. The setup for measuring tensile strength by the split cylinder test is depicted. Notice that the load is applied normal to the length rather than normal to the cross section, as is done in measurements of compressive strength. *Source*: Sharifinassab, M. An experimental investigation of concrete and limestone deterioration by crystallization of sodium sulfate, M.S. thesis, University of Louisville, 1981, p. 27.

where σ_T is the tensile strength, P is the load at failure and D and L are the diameter and the length of the sample, respectively.

Tables 3.1 and 3.2 give, respectively, the data and the average compressive and tensile strength, each measured on two samples, as well as the calculated reduction in strength resulting from salt crystallization in each cycle. A rapid reduction in compressive strength and tensile strength is evident. This change may be used to estimate the rate of deterioration. Henceforth, we will call the deterioration by tensile force as disintegration and that by compressive force as breakdown of cohesion and represent these by R_d and R_b, respectively.

Standard relationships exist to determine the engineering strengths in compression, σ_C [Eq. (3.4)], and tension, σ_T [Eq. (3.5)], which suggest that the reduction in strength is proportional to the decrease in force (P) required to produce failure in each case.

$$\sigma_C = \frac{P}{\pi r^2} \tag{3.4}$$

$$\sigma_T = \frac{P}{\pi r L} \tag{3.5}$$

TABLE 3.1 Data for Compressive Strength and Computed Average Strength and Percent Reduction in Cycles of Sodium Sulfate Crystallization

Cycle	Load (kg)	Length (cm)	Diameter (cm)	Compressive Strength (kg/cm^2)	Compressive Strength	
					Avg. (kg/cm^2)	Reduction (%)
0	4041	8.00	3.81	356.77		
	4585	7.62	3.81	401.76	379.26	
I	3587	8.13	3.81	314.24		
	3632	7.87	3.81	318.25	316.24	17
II	2951	8.25	3.81	258.56		
	3178	8.13	3.81	278.45	268.51	29
III	2406	7.37	3.81	210.90		
	2429	7.62	3.81	212.87	211.88	44

Source: Sharifinassab, M. An experimental investigation of concrete and limestone deterioration by crystallization of sodium sulfate, M.S. thesis, University of Louisville, p. 66, 1981.

If it is assumed that the strength of the virgin material is unchanged from sample to sample, then Eq. (3.5) for tensile strength may be used to compute material deterioration due to crystallization effect. The tensile strength is used because the depth of microcracking, which can be assumed to occur by salt crystallization, tends to influence tensile strength more readily than it will influence compressive strength. Therefore, using Eq. (3.5) the deterioration (R_d) can be given by

TABLE 3.2 Data for Tensile Strength and Computed Average Strength and Percent Reduction in Cycles of Sodium Sulfate Crystallization

Cycles	Load (kg)	Length (cm)	Diameter (cm)	Tensile Strength (kg/cm^2)	Tensile Strength	
					Avg. (kg/cm^2)	Reduction (%)
0	2542	7.62	3.81	55.68		
	2270	7.87	3.81	48.08	51.88	
I	1589	8.25	3.81	32.12		
	1839	8.13	3.81	37.75	34.94	33
II	1135	7.62	3.81	24.89		
	953	7.87	3.81	20.25	22.49	57
III	431	8.25	3.81	8.72		
	477	7.64	3.81	10.40	9.56	82

Source: Sharifinassab, M. An experimental investigation of concrete and limestone deterioration by crystallization of sodium sulfate, M.S. thesis, University of Louisville, p. 68, 1981.

$$R_d = \frac{P - P_n}{\pi \sigma_T L} \tag{3.6}$$

where P is the force on virgin material producing the original tensile strength, P_n is that force after the nth cycle of crystallization, and σ_T is the original tensile strength. For example, after the first cycle, averaging the load on two samples (Table 3.2),

$$R_d = \frac{2406 - 1714}{3.14 \times 51.88 \times 7.75} = 0.55 \text{ cm}$$

Values of R_d for various cycles (Table 3.3), ignoring the somewhat aberrant value in third cycle, give 0.55 cm as the *rate* of disintegration (r_d) per cycle.

The compression test, as indicated above, may be viewed as a measure of the effective rate of decrease of cohesion in the material. This decrease is less than the deterioration in tension and signifies the depth of loss of cohesion of the material. The relationship for compressive strength according to Eq. (3.7) is used to determine the depth of cohesion breakdown (R_b),

$$R_b = \frac{\sqrt{P} - \sqrt{P_n}}{\sqrt{\pi \sigma_C}} \tag{3.7}$$

where σ_C is the original compressive strength of the material.

Thus, after the first cycle, the effected thickness is

$$\frac{\sqrt{4313} - \sqrt{3609}}{\sqrt{3.14 \times 379}} = 0.16 \text{ cm}$$

Values of R_b in various cycles (Table 3.3) give nearly 0.18 cm as the *rate* of breakdown (r_b) per cycle. This is about one-third as much as the rate of deterioration (r_d) measuring the extent of microcracking in the material.

To determine the extent of the depth to the which the salt had been deposited in the stone, measured increments of cylindrical specimen were shaved off after one cycle of crystallization. Shore hardness (see Sec. 3.4) measurements were made on exposed surface after shaving off a layer. Table 3.4 gives essential points showing the resulting change in hardness as a function of depth (Fig. 3.17). The data suggest that hardness

TABLE 3.3 Disintegration (R_d) and Depth of Cohesion Breakdown (R_b) Calculated after Eqs. (3.6) and (3.7)

Sample	R_d (cm)	R_b (cm)
Cycle I	0.55	0.16
Cycle II	1.08	0.30
Cycle III	1.11	0.48

TABLE 3.4 Depth of Salt Penetration (One Immersion) in Stone Measured as a Function of Shore Hardness[a]

Depth from Surface (mm)	Shore Hardness
0.000	20.0
0.160	21.0
0.684	21.0
1.836 (1)	24.5
3.179	26.0
4.020	25.5
5.188	26.5
7.120 (2)	27.0
8.186	27.5

[a] See Figure 3.17; (1) and (2) indicate depths at which rapid change in shore hardness is noticed.

is affected up to about 7 mm depth, which somewhat agrees with the rate of deterioration per cycle (r_d). The rate of material breakdown (r_b) also appears to be concurrent with the material depth at which the rate of decrease of hardness is greatest, that is, about 1.6 mm. These values are marked as (2) and (1) in Table 3.4.

3.3.2 Modulus of Elasticity

In elastic deformation, that is, deformation that is recovered if the stress is removed, represented in a stress–strain graph (Fig. 3.3) for compression or tension, the slope equals the stress-to-strain ratio. This ratio is called the modulus of elasticity (E) or Young's modulus. Thus, the modulus of elasticity expresses materials resistance to deformation. Commonly, modulus is used for selection of new materials for construction or for replacement of stone that has deteriorated in existing structures. We will use the change in modulus in increasing cycles of salt crystallization as a measure of the deterioration of stone.

Tangent modulus is commonly used taken at a stress level equal to one-half the ultimate strength (Fig. 3.18). Thus, E for the virgin stone is unit stress (kg/cm^2)/unit strain (cm/cm) = $(185 - 141)/[(7.33 - 6.00) \times 10^{-4}] = 33.10 \times 10^4$ kg/cm^2. Accordingly, the modulus for other samples was calculated and is shown in Table 3.5.

Now, we can predict a change in the modulus of elasticity, E_0/E_n, after n cycles of sodium sulfate crystallization as follows:

$$E_0 = \frac{P/\pi r_0^2}{\Delta l/l_0} \quad \text{and} \quad E_n = \frac{P/\pi r_n^2}{\Delta l/l_n} \tag{3.8}$$

Thus, assuming l_0 and l_n are the same,

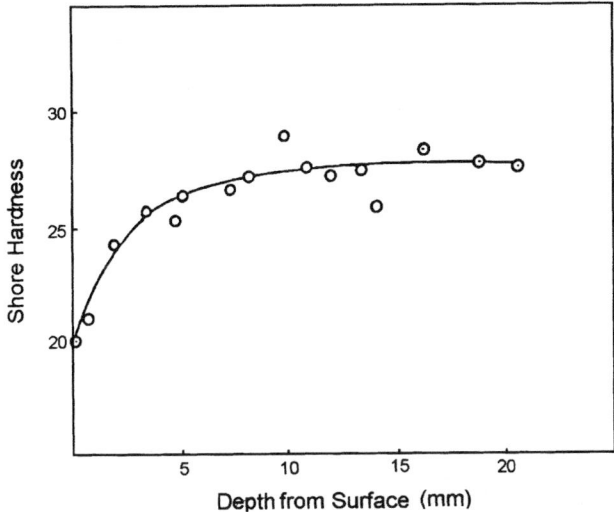

FIGURE 3.17 Depth of sodium sulfate penetration measured by the change in Shore hardness of surfaces prepared by shaving a known thickness of core. *Source:* Cassaro, M. A., Gauri, K. L., Sharifinassab, M. and Sharifian, A. in K. L. Gauri and J. A. Gwinn (eds.), *On the Strength and Deformation Properties of Indiana Limestone and Concrete in the Presence of Salts*, The University of Louisville, 1982, pp. 57–76.

$$\frac{E_0}{E_n} = \frac{r_n^2 l_0}{r_0^2 l_n} \tag{3.9a}$$

$$E_n = \frac{E_0 r_0^2}{r_n^2} \tag{3.9b}$$

The following example, based upon our knowledge of the original modulus (33.10×10^4) and the rate of breakdown in compression (0.15 cm/cycle), gives the value of a new modulus after one cycle of crystallization for a sample of 3.81 cm radius:

$$E_1 = 33.10 \times 10^4 \frac{(3.81 - 0.15)^2}{(3.81)^2} = 30.54 \times 10^4 \text{ kg/cm}^2$$

This value has the same order of magnitude as the experimentally obtained values given in Table 3.5.

The importance of the preceding technique to stone weathering and conservation can be seen in a number of ways. For example, the modulus of elasticity of sound Indiana limestone being known and the depth of disintegration of the material measured on cores obtained from a weathered structure, the new modulus of elasticity

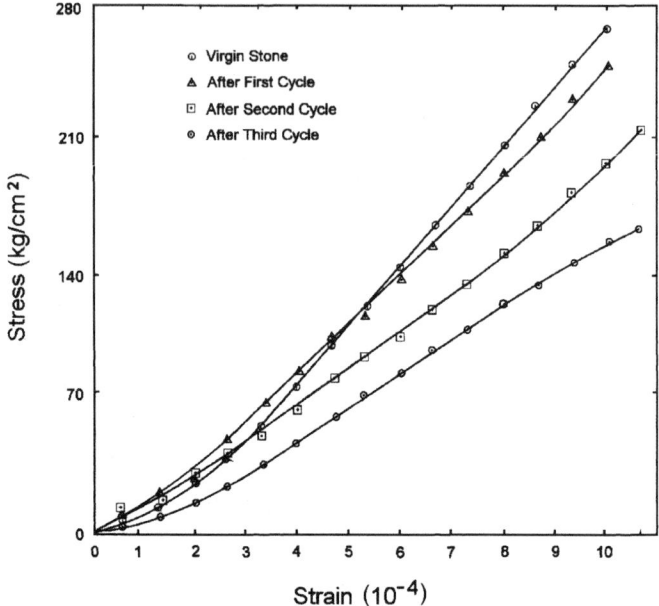

FIGURE 3.18 Stress–strain curves in the limit of elastic strain. Shown are relationships for Indiana limestone before sodium sulfate crystallization and after three cycles. *Source*: Cassaro, M. A., Gauri, K. L., Sharifinassab, M. and Sharifian, A. in K. L. Gauri and J. A. Gwinn (eds.), *On the Strength and Deformation Properties of Indiana Limestone and Concrete in the Presence of Salts*, The University of Louisville, 1982, pp. 57–76.

TABLE 3.5 Stress (kg/cm²) at Two Adjacent Points Toward the Middle of the Stress–Strain Curve in Elastic Deformation and the Corresponding Strain (cm/cm), and the Calculated Modulus of Elasticity (E)

Sample	Stress (kg/cm²)	Strain (10^{-4} $\delta l/L$)	Modulus of Elasticity E (10^4 kg/cm²)
Virgin Stone	141	6.00	33.10
	185	7.33	
Cycle I	136	6.00	26.31
	171	7.33	
Cycle II	105	6.00	21.80
	134	7.33	
Cycle III	85	6.00	18.05
	109	7.33	

can be readily calculated, giving the magnitude of change due to weathering. Alternatively, if this stone is consolidated (Chap. 10), the desired improvement in the modulus of elasticity can be made as a specification for the prescribed treatment.

3.4   COMPRESSIVE STRENGTH OF STONE WITH CRUST

Carbonate rocks in outdoor exposure develop a surface crust when protected from rain (Chap. 5). Beneath the crust is the leached layer that supplied materials that formed the crust. The crust and the leached layer constitute the zone of weathering, which is often a very thin appendage upon the fresh stone. Therefore, the compressive strength of weathered zone cannot be determined by engineering techniques. We will describe an indirect method through which changes in the compressive strength within the zone of weathering can be measured.

Miller (1965) investigated certain index properties used to predict physical behavior of rocks. These properties include Schmidt hardness, Shore scleroscope hardness, and sonic wave velocity. Schmidt hardness being based on high impact energy (0.54 foot-pounds = 0.8 m·kg) is too insensitive for investigation of weathered rocks. For measurement of sonic velocity, large homogeneous specimens are required. However, the Shore scleroscope hardness in conjunction with unit weight worked out well for use in strength calculations.

A Shore scleroscope is a nondestructive hardness-measuring device, which indicates values of hardness by rebound of a small diamond-pointed hammer dropped vertically onto a test surface. A Model D Shore scleroscope (Fig. 3.19) was found to be sufficiently sensitive for use on carbonate rocks.

Experimental data on uniaxial compressive strength, Shore hardness, and apparent specific gravity of limestone, published through the United States Bureau of Mines (USBM) by Blair (1955, 1956), Obert et al. (1946), and Windes (1949, 1950) best fit an equation of the form

$$\ln \sigma = A + B \, (\text{ASG}) \, (S) \tag{3.10}$$

where, σ is the uniaxial compressive strength; ASG is the apparent specific gravity; S is the Shore scleroscope hardness; and A and B are constants, with the value of 0.914 and 0.014, respectively.

To obtain the curve for Shore hardness versus depth into stone (Fig. 3.17), the core is successively shaved from the weathered surface inwards until unweathered marble is reached. At each level a number of Shore scleroscope hardness (S) values are obtained and the apparent specific gravity (ASG) calculated from the weight and volume of the shaved portion (see Sec. 5.2.3 and Appendix C for measuring porosity). The values of S and ASG at any depth within the zone of weathering divided by corresponding values of fresh stone, that is, the stone beneath the zone of weathering, give normalized values of S and ASG. The depth (D_n) is normalized by dividing the depth from surface with the thickness of the zone of weathering. The method to normalize the S, ASG, and D is illustrated by Fig. 3.20. For example, the average Shore hardness at 0.1 mm is 30, ASG for the weathered layer up to 0.1 mm is 2.0,

FIGURE 3.19 Shore scelerscope model D. The manufacturer's standard samples are used to calibrate the instrument. Notice a marble sample showing a Shore hardness of 39.

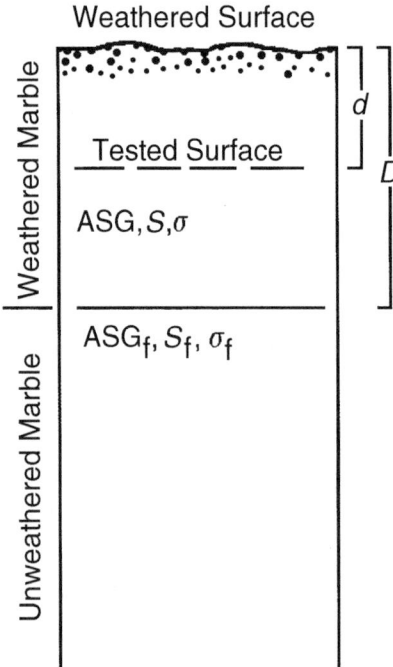

FIGURE 3.20 Weathered marble with crust. The normalization protocol for Shore hardness (S), apparent specific gravity (ASG), and compressive strength (σ) are indicated. S_f, ASG_f, and σ_f are in the region beyond the zone of weathering. To normalize data, S, ASG, and σ at known depth in the zone of weathering is divided by corresponding value of fresh marble. The depth is normalized (D_n) by dividing a given depth in the zone of weathering (d) by the thickness (D) of the zone of weathering. *Source:* Gauri, K. L. et al., *Eng. Geol.* **6:** 235, 1972.

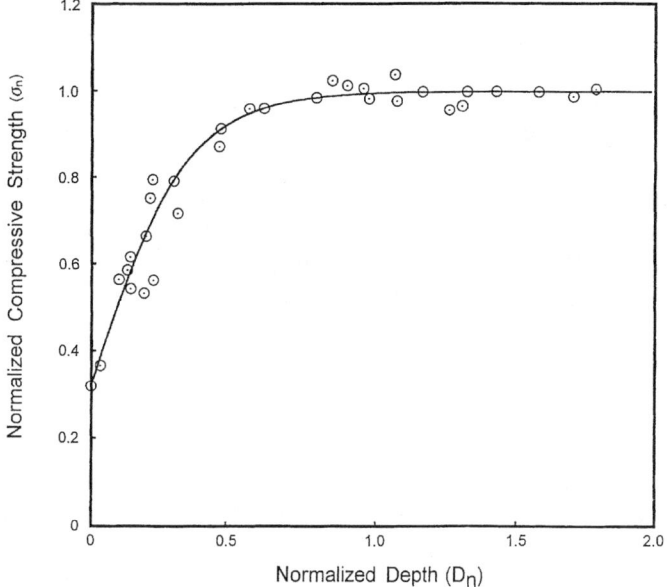

FIGURE 3.21 Change of the compressive strength of weathered marble with crust. Normalized data of a marble core cut from a 100-year old tombstone which had weathered at a site protected from rain in Cave Hill Cemetery in Louisville. *Source:* Gauri, K. L. et al., *Eng. Geol.* **6:** 235, 1972.

thickness of the zone of weathering is 0.2 mm, and the S and ASG of nonweathered marble is 40 and 2.8, respectively. Then the D_n at 0.1 mm is 0.5 (0.1 mm/0.2 mm), ASG_n is 0.7 (2/2.8), and S_n is 0.75 (30/40). In Figure 3.20, S_f, ASG_f, and D_f represent the upper surface of nonweathered marble.

Figure 3.21 shows the normalized compressive strength of a core obtained from a nearly 100-year-old weathered marble monument. Now, the curve depicting normalized compressive strength (σ_n) can be expressed by:

$$\ln(\sigma_n) = -1.148 \exp(-5.5\, D_n) \tag{3.11}$$

Accordingly, σ_n in the zone of weathering ranges from 0.31 to 0.99. Thus, the compressive strength is reduced up to 31%. Assuming that the nonweathered marble has a compressive strength of 700 kg/cm^2, the approximate compressive strength of the zone of weathering ranges from 220 kg/cm^2 to 700 kg/cm^2.

SUMMARY

Carbonate rocks are brittle materials; therefore, they commonly contain joints. The study of the orientation of joints is useful in elucidating the weathering behavior of these rocks. Methods to measure joints and their representation in stereographic projections are explained.

The mechanical strength of the material changes with weathering. Concepts of geomechanics, including compressive strength, tensile strength, and modulus of elasticity, are applied to determine the rate of deterioration from internal pressure developed by salt crystallization.

Many materials develop a narrow zone of weathering so that their strength properties cannot be determined by routine engineering tests. We have described a method whereby the compressive strength of thin layers within the zone of weathering can be determined from hardness and specific gravity.

REFERENCES

Blair, B. E., Physical properties of mine rocks, 2, *U.S., Bur. Mines, Rep. Invest.*, 5130, 1955.

Blair, B. E., Physical properties of mine rocks, 4, U.S., *Bur. Mines, Rep. Invest.*, 5244, 1956.

Miller, R. P., Engineering classification and index properties of intact rock, thesis, University of Illinois, Urbana, Ill., 1965.

Obert, L., Windes, S. L., and Duvall, W. I., Standardized test for determining the physical properties of mine rocks, *U.S., Bur. Mines, Rep. Invest.*, 3891, 1946.

Windes, S. L., Physical properties of mine rocks, 1, *U.S., Bur. Mines, Rep. Invest.*, 4459, 1949.

Windes, S. L., Physical properties of mine rocks, 1, *U.S., Bur. Mines, Rep. Invest.*, 4727, 1950.

CHAPTER 4

Weathering of Carbonate Rocks in Natural Environments

4.1 INTRODUCTION

Weathering includes processes, such as chemical action of air and rainwater and the mechanical action of water, whereby rocks when exposed to atmosphere decay into new products or disintegrate into fragments.

A natural environment, also called clean air, is considered to be one that prevails currently in nonindustrial countries and also one that prevailed in modern industrial countries before the industrial revolution. One characteristic of the natural environment pertinent to rock weathering is that while it contains the chemically active gas carbon dioxide (CO_2), also found in polluted air, it lacks the industrial affluent sulfur dioxide (SO_2) and nitrogen dioxide (NO_2).

Carbonate minerals do not react with carbon dioxide in the gas phase but are slightly soluble in rainwater containing dissolved carbon dioxide. However, the oxides of sulfur and nitrogen are strongly reactive in both gas and liquid phases. Thus monuments that currently exist in natural environments, or those that were removed from outdoor exposure before air pollution set in, are chemically well-preserved after even millennia of outdoor exposure. In contrast, many of those exposed even for a few decades in industrial environments have decayed beyond recognition.

Weathering also includes disintegration by forces generated within the rock. Because carbonate rocks contain a network of pores and cracks, aqueous solutions can penetrate into rock and damage it in a variety of ways including loss of cohesiveness between grains by wetting and bursting pressure from freezing, salt crystallization, and swelling clay minerals, if present. These are the processes of mechanical weathering, which are also fully operative in polluted environments. They are discussed in detail in this chapter because we believe that the primary cause of weathering in a natural environment is mechanical rather than chemical weathering.

4.2   CHEMICAL WEATHERING IN A NATURAL ENVIRONMENT

Carbonate rocks are made of the minerals calcite ($CaCO_3$) and dolomite [$CaMg(CO_3)_2$]. In a natural environment, carbon dioxide is the only gas that reacts with these minerals. However, metallic objects are sometimes used in the construction of masonry structures. For example, iron anchors are commonly applied to fasten dimensional blocks and sculpted stone to the structural framework. The oxygen in the atmosphere, which is nonreactive with carbonate minerals, reacts with iron and forms iron oxides, which then disrupt the carbonate rock. We will discuss below the chemistry of dissolution of carbonate rock and the iron-oxidation reactions later.

4.2.1   Solution of Carbonate Minerals

Carbonate minerals dissolve in rainwater containing carbon dioxide following the reactions:

$$CaCO_3 + H_2O + CO_2 \rightarrow Ca^{2+} + 2HCO_3^- \qquad (4.1)$$

$$CaMg(CO_3)_2 + 2H_2O + 2CO_2 \rightarrow Ca^{2+} + Mg^{2+} + 4HCO_3^- \qquad (4.2)$$

Dissolved calcium (Ca^{2+}), magnesium (Mg^{2+}), and bicarbonate (HCO_3)$^-$ ions are drained away, causing surface erosion of monuments in a short historical time frame and producing caverns in the body of the rock over geologic time.

In limestone terrain these reactions have produced karst landforms characterized by the presence of underground caves, such as Mammoth Cave in Kentucky. The conditions for massive removal of material were, in addition to long time, highly fractured rock, which allowed passage of a large volume of underground water, and underground water saturated with carbon dioxide from soil, the concentration of which is orders of magnitude higher than atmospheric carbon dioxide. The same reactions hold for the weathering of carbonate monuments in natural atmosphere, but most monuments show only slight decay even when exposed outdoors for millennia. The reasons, in contrast to those indicated for caves, namely, passage of a large volume of water through fractures over long geologic time and high CO_2 concentration of the underground water, are as follows:

- Monuments and historic structures are often protected from rain. Because carbon dioxide gas does not react with carbonate minerals in the absence of liquid water, these objects are not affected.
- Rain in equilibrium with modern atmospheric carbon dioxide of nearly 360 ppm produces a rainwater of pH 5.6. Compared with pure water, which has a pH of 7.5, modern clean rain is slightly acidic. However, in pre-industrial environment, which can be said to have existed up to 1850, the carbon-dioxide composition was as low as 280 ppm (Andrew et al., 1996, p. 163). Thus, the

rains were then even less acidic and their capacity to dissolve carbonate minerals lower.

We will now illustrate by a few examples that the chemical weathering of historic monuments in nonindustrialized environments and in the modern industrialized environments prior to their pollution by sulfur and nitrogen oxides, had indeed been insignificant.

- The casing limestone blocks at the apex of the nearly 5,000-year-old Khufu Pyramid in Egypt still show the original whitish limestone surface.
- A nearly 2,500-year-old Caryatid (Fig. 4.1) moved from the modern polluted air of Acropolis of Athens into the British Museum in 1816 does not show any fading of relief.

FIGURE 4.1 Comparison of chemical weathering in natural and polluted environments. A natural environment is one in which carbon dioxide is the major reactive gas and oxides of sulfur and nitrogen, products of industrial processes, are absent. Caryatids from the temple of Erechtheum at the Acropolis, Athens, Greece, are shown. The Caryatid on the left was removed from the Erechtheum and placed in the British Museum, London, in 1816, while the Caryatid on the right was reinstalled in early 1900s during the last restoration of Acropolis. It is evident that the Caryatid protected from the modern atmosphere of Athens hardly shows any effects of chemical weathering despite the fact that it was exposed to natural environment for over two millennia before it was moved to the relatively clean environment of the museum.

- Both rain-exposed and rain-protected marble surfaces at the nearly 350-year-old Taj Mahal in India are chemically well-preserved (Gauri and Holdren, 1981).

We have indicated that chemical weathering of monuments consists of the breakdown of the constituent minerals, which are then drained away. The result is loss of surface relief, often termed surface reduction, whereby features and inscriptions become faint over time. However, in considering the magnitude of reduction, often calculated from the leached ionic mass (Chap. 6), a point often ignored but significant is the contribution of calcium and carbonate ions transported from the material's depth. Given suitable conditions for a porous carbonate rock to remain wet for an extended period of time, a supersaturated solution of calcium carbonate can be produced in the pores. Drying at the surface draws this solution outward, where recrystallization of calcite can occur. Thus, the near-surface pores become partially plugged and a crust forms on the surface (Fig. 4.2). The crust actually marks a gain of surface rather than reduction, although some soluble ions are drained away. Thus, the reduction of the original surface—calculated from ions in the rain runoff—may often be from the mass that had been gained by the surface from within the pores.

4.2.2   Oxidation of Iron Anchors

Often, damage to historic structures occurs through oxidation of iron artifacts inserted in the stone blocks for anchoring. Water is ordinarily saturated with atmospheric oxygen. Reaction of oxygen with iron occurs through an electrochemical process. If a water film surrounds an iron object, iron is oxidized at the anodic area to Fe^{2+} ions. At the cathodic area, elementary oxygen is reduced to OH^- ions. The Fe^{2+} ions move through the water film from anode to cathode. At the cathode, they react with OH^- ions to produce $Fe(OH)_2$, which is then oxidized to $Fe(OH)_3$. The following equations express these reactions:

$$(+)Fe \rightarrow Fe^{2+} + 2e^- \tag{4.3a}$$

$$(-)\tfrac{1}{2}O_2 + H_2O + 2e^- \rightarrow 2OH^- \tag{4.3b}$$

$$2Fe(s) + O_2(g) + 2H_2O \rightarrow 2Fe(OH)_2(s) \tag{4.3c}$$

$$2Fe(OH)_2(s) + \tfrac{1}{2}O_2(g) + H_2O \rightarrow 2Fe(OH)_3(s) \tag{4.3d}$$

The final reddish brown product is limonite [$Fe(OH)_3$], the flaky deposit called rust. This occupies more than twice the volume of the parent iron; its density is less than 3 while that of iron is 7.8. The pressure generated in the confined space of the insert by the volumetric increase is often large enough to fragment the stone. This type of

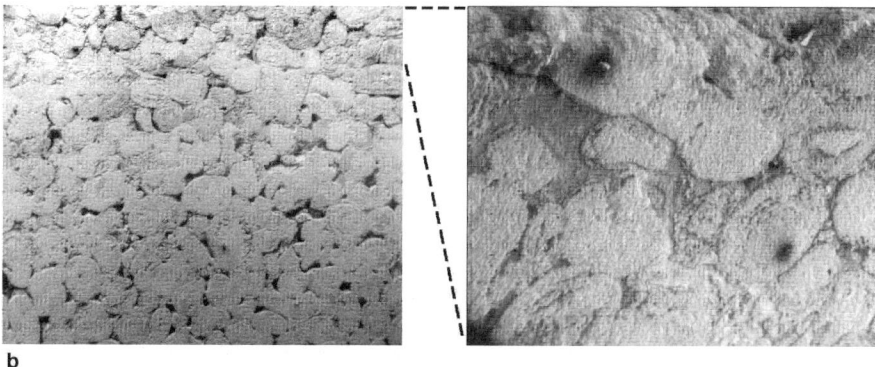

FIGURE 4.2 Recrystallization of calcite. If water is allowed to stay in pores for some time, it becomes saturated with calcium and bicarbonate ions. On drying, calcium carbonate is precipitated. (a) A recrystallized calcite crust is shown; it is white due to the leaching of iron present in the rock. (b) Scanning electron micrographs of a transverse section of a sample with crust: (left) pores nearly plugged with calcite close to the exterior surface (top), but progressively empty farther away; (right) upper portion enlarged to show the pores filled with recrystallized calcite. The rounded structures with concentric shells are oolites, which make up this limestone. *Source*: Gauri, K.L., The Preservation of Stone, *Scientific American*, **238** (6): 126–136, 1978.

weathering may be considered as a mechanical effect of chemical decay. We will give two examples to illustrate significance of this phenomenon in historic architecture.

The Acropolis of Athens is the site of many temples constructed in marble nearly 2,500 years ago. Over the years, temples were dismantled and stone blocks scattered or used in construction of buildings many of which had been built over the ancient monuments. The last major restoration of the temples at Acropolis occurred between

a

b

FIGURE 4.3 Disintegration of marble by oxidation of iron anchors. (a) Temple of Erechtheum restored in the late nineteenth and early twentieth centuries using iron anchors, which have since oxidized and disintegrated the marble blocks seen clearly along joints at rear-right. In the original construction of the temple in the fifth century B.C., iron anchors were coated with lead that inhibited oxidation. (b) Taj Mahal, constructed from 1631 to 1653; a section of the mausoleum showing an iron dowel and the disintegrated marble surrounding the dowel. Notice the near-perfect preservation of sculptures in marble where not affected by iron oxidation.

1886 and 1933 under the stewardship of the architect Nicolas Balanos. To restore the structures, steel elements were employed. Iron anchors had been employed by the ancient Greeks in the construction of these monuments. To prevent rusting, however, they coated the anchors with lead, a practice almost universally followed until recently. Balanos perhaps overestimated the corrosion-resistant properties of the steel. In modern architecture, even stainless steel, professed to be nonreactive, begins to rust after a few years of exposure. The oxidation of steel anchors applied in the restoration by Balanos resulted in the disintegration of marble in the vicinity of the anchors [Fig. 4.3(a)], compounding the deleterious effects of air pollution (Chap. 5).

Taj Mahal, erected between 1631 and 1653, illustrates another example of marble disintegration [Fig. 4.3(b)] by similar processes. The fine architectural features of the Taj Mahal include delicate carvings on massive marble blocks. At present, most carvings are extremely well preserved because of the lack of air pollutants, but the disruption by oxidation has necessitated replacement of many stone blocks by new marble.

4.3 PHYSICAL WEATHERING

Physical weathering, also called mechanical weathering, is the disintegration of material into smaller particles, which then can be removed by gravity, wind, water, or ice. These erosional processes further result in the exposure of fresh rock to weathering.

Physical weathering depends upon the pores and fractures in the rock through which water can penetrate the rock mass. Once inside, water can cause disintegration of the rock by wetting, expansion on freezing, crystallization of salts on drying, and expansion of clay minerals many of which readily adsorb water.

4.3.1 Wetting

Wetting is the adsorption of water on pore walls. When adsorbed, water reduces the mechanical strength of rock in two ways. First, the rock may expand. The linear expansion of Indiana limestone on wetting, for example, is 0.006%. Cyclical wetting and drying will fatigue the rock, causing it to rupture. Second, water reduces cohesiveness between grains by disrupting intermolecular forces. For example, the ratio of the uniaxial compressive strength of water-saturated to oven-dried specimens of Indiana limestone is 0.59. In fact, the strength of rock with water adsorption is a continuous function of water content but is considerably reduced when the water content is as low as 5% of saturation content. This indicates that moisture, adsorbed at prevailing humidity and then condensed in fine capillaries, can cause a considerable reduction in the strength of rock. We will discuss the subject of condensation of moisture in the context of salt crystallization later. Here, we elaborate the mechanism through which rainwater or underground water is able to intrude masonry structure.

4.3.1.1 Mechanism of the Intrusion of Water into Masonry A   film   of water consists of many molecules that are attracted to each other through intermolecular forces. The molecules in the interior of the film are subject to a greater attractive force than the molecules at the surface, because the interior has a larger number of surrounding molecules than the surface, in which molecules have fewer neighbors. The internal intermolecular force gives cohesiveness to the water film, while a lesser attractive force at the surface produces surface tension. The surface tension of water at 20°C is 7.28×10^{-2} newtons per meter (N·m^{-1}). This force must be exceeded by the force in the pores, called capillary attraction, in order for water to enter the masonry structure.

All liquids tend to achieve as small a surface area as possible because of the attractive intermolecular forces existing within the liquid. Thus, when freely suspended, liquids tend to become spherical drops, a configuration that represents the smallest surface-to-volume ratio. On contact with a solid, the surface of a drop of liquid forms an angle with respect to the solid surface called the contact angle (Fig. 4.4). The contact angle is determined by competition between the cohesive liquid–liquid molecular forces and the adhesive solid–liquid forces and depends on the particular solid and liquid involved. Hydrophilic materials, that is, materials that attract water, elongate or pull the drop outwards, resulting in a small contact angle less

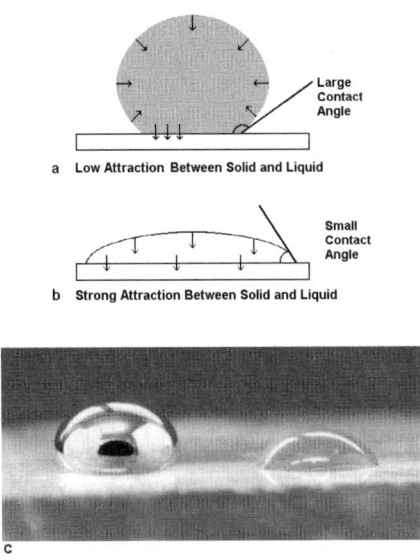

a   Low Attraction Between Solid and Liquid

b   Strong Attraction Between Solid and Liquid

c

FIGURE 4.4   Contact angle. The contact angle is the angle that surface of a drop of a liquid makes with the substrate and represents the interaction between a solid and a liquid. (a) A large contact angle is indicative of a nonwetting, or hydrophobic, material, while (b) a small contact angle represents hydrophilic material. (c) shows drops of (left) mercury, a nonwetting liquid, and (right) water, a wetting liquid, on the surface of a calcite crystal.

than or equal to 90°, whereas hydrophobic materials have a contact angle larger than 90°.

Carbonate rocks are hydrophilic materials. Intrusion of water into masonry occurs because water, though electrically neutral, is a polar molecule with a net positive charge in one direction and that carbonate minerals contain oxygen which is electronegative. Thus, water molecules are attracted by the masonry structure (Fig. 4.5) and pulled along the pore walls by capillary force, also referred to as capillarity or capillary action.

In a capillary, a hairlike pore, the maximum electrochemical force occurs at the walls and decreases towards the center. The resultant surface of the water table, called the meniscus, is therefore concave upwards. In a wider pore filled with water, the large cohesive force in the water tends to pull the water away from the pore wall. Thus, the meniscus in a large pore can be nearly flat (Fig. 4.6). This shows that the radius of a capillary determines the rate of transport of water and the height to which water can rise; small pores are filled more quickly and to a greater height than are large pores. This knowledge leads to the understanding of the following sequence in which pores of different sizes in a rock will be filled (Fig. 4.7):

- Surfaces of capillaries adsorb water first until all capillaries are filled. The large pores are dry.
- Surfaces of large pores adsorb water from capillaries.
- Large pores are filled.

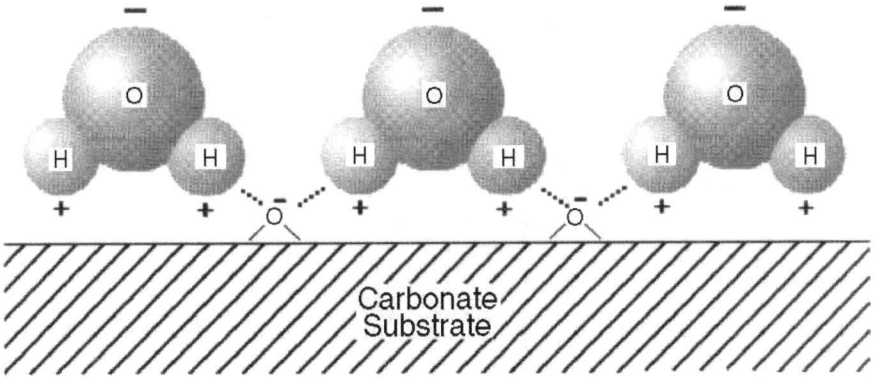

FIGURE 4.5 Attraction of water by a carbonate surface. Water is a polar molecule with the positive and negative charges oriented preferentially, and the carbonate surface is electronegative because of the excess negative charge upon oxygen. Therefore, carbonate surfaces attract water. *Source*: Torraca, G., *Porous Building Materials. Materials Science for Architectural Conservation*, 3rd ed., International Center for the Study of Preservation and the Restoration of Cultural Property, ICCROM, p. 5, 1988; courtesy of ICCROM.

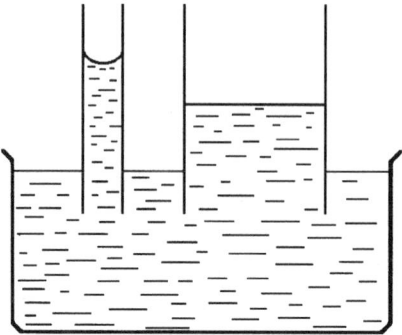

FIGURE 4.6   Control of capillary radius upon the rate of water suction and height of water column. The upper surface of a water column in a narrow capillary, called the meniscus, is concave upwards because of the attraction of water by the capillary wall. Intermolecular forces, however, pull the water molecules towards the center of the water column. Therefore, in a capillary with large radius the meniscus can be flat. The rate of rise of water is greater and the height to which the water will rise is higher in a fine capillary.

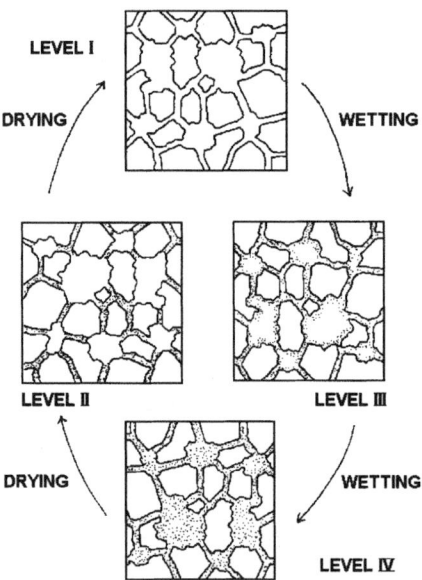

FIGURE 4.7   Sequence of the filling of small and large pores. Following the description of Figure 4.6, small pores are filled first and then supply water to the large pores. Thus, depending upon the availability of water, some large pores may not be filled at all. The sequence of drying is in the reverse order of filling. *Source*: Torraca, G., *Porous Building Materials. Materials Science for Architectural Conservation*, 3rd ed., International Center for the Study of Preservation and the Restoration of Cultural Property, ICCROM, p. 10, 1988; courtesy of ICCROM.

This sequence is reversed when the rock dries. The large pores dry first and capillaries last.

We now look at the relationship of contact angle, surface tension, gravity, and radius of pore, which determines the height to which water will rise in a pore. Figure 4.8 shows that the net vertical force F_{up} is the vertical component of the surface tension times the length l of the water surface in contact with the pore; l equals the circumference $2\pi r$. The radius of the pore is r and ρ is the density of water. Thus

$$F_{up} = 2\pi r \gamma \cos \theta \tag{4.4}$$

The volume of the column of water from the bottom of the pore to the bottom of the curved water surface or meniscus is $V = \pi r^2 h$, and its weight is $w = \rho g V = \rho g \pi r^2 h$. The liquid rises until $F_{up} = w$, or

$$2\pi r \gamma \cos \theta = \rho g \pi r^2 h \tag{4.5}$$

Thus

$$h = \frac{2\gamma \cos \theta}{\rho g r} \tag{4.6}$$

For example, water rises in a capillary of radius $r = 2.5 \times 10^{-5}$ m. The contact angle is $0°$. The density of water is 10^3 kg·m^{-3} and the surface tension γ at 20°C is 7.28×10^{-2} N·m^{-1}; thus

$$h = \frac{2(7.28 \times 10^{-2} \text{N·m}^{-1})}{(10^{-3} \text{ kg·m}^{-3})(9.8 \text{ m·s}^{-2})(2.5 \times 10^{-5} \text{ m})} = 0.594 \text{ m}$$

Water can enter the pores not only vertically as discussed previously, but it can also migrate laterally. The lateral rate of transport can be given by the simple expression (Amoroso and Fassina, 1983, p. 17):

$$x = A\sqrt{t} \tag{4.7}$$

where x is the distance (m) over which water will travel in time t (s) and A is a constant that can be determined experimentally. Also, A relates to the capillary radius (r), the surface tension γ (N/m), and dynamic viscosity η (N·s/m^2) and can be calculated according to the expression:

$$A = \sqrt{\frac{\gamma r}{2\eta}} \tag{4.8}$$

These relationships hold true for homogenous materials with cylindrical pores, whereas rocks are not homogenous nor do they have pores that are cylindrical. Furthermore, it is desirable to know the rate of movement of water into stone in

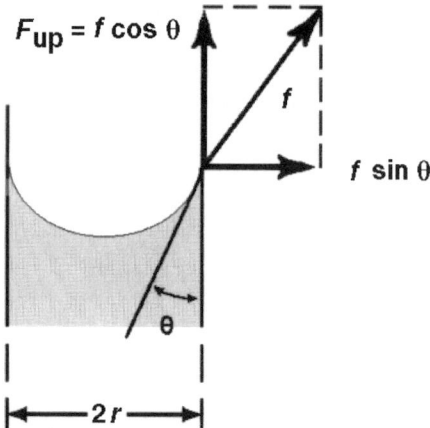

FIGURE 4.8   Vertical rise of liquid in a capillary. Liquid in a capillary of radius r with a contact angle θ rises to a height h. The force f on a small segment of the liquid in contact with the capillary is shown. *Source*: Kane, J. W. and Sternheim, M. M., *Physics*, 2nd ed., New York: Wiley, p. 306. Copyright 1983. Reprinted by permission of John Wiley & Sons.

addition to the ultimate depth or height to which water will penetrate into rock. To obtain an idea of the amount of wetness over time, experiments can be carried out for each rock in question. Figure 4.9 shows, as examples, the intake of water in various limestones, dolomites, and marbles described in several chapters in this book.

4.3.1.2 Capillary Rise and Pore Size

Carbonate rocks contain pores of different sizes; however, those with small radii are effective in the rise of water into stone. In this section we will give a method of calculation of mean pore radii of small pores in a rock system.

The flow of fluid through capillary depends upon the pressure head (P), kinetic energy (v), hydraulic head (H), frictional shear force (h_f), and force due to surface tension (F_T) of the liquid. The total momentum balance equation for fluid flow through capillary is given by

$$\frac{dP}{\rho} + \frac{dv^2}{2g_c} + \frac{g}{g_c}\,dH + dh_f + F_T = 0 \tag{4.9}$$

If the capillary is open at both ends, then $dP = 0$. Furthermore, as the velocity of fluid in the capillary is very small, the kinetic energy contribution can be neglected. Therefore the main contributing factors are the frictional wall shear force and the force due to surface tension of the fluid. For laminar flow regime

a

b

c

FIGURE 4.9 Capillary rise of water. The samples were brought in contact with water over time in a Petri dish and covered with beakers to avoid evaporation. (a) Limestone (from left to right): Cordova Cream, Cordova Shell, Indiana limestone, and Lueders; 4 h. Notice that in Cordova Cream the water has risen nearly to the top of the sample, whereas in Indiana limestone the water has gone up less than one-third this height. (b) Dolostone: Big Blue (left); Laurel dolomite (right); 2 h. (c) Marble: Carrara marble (left); Georgia marble (right); 2 h. The height of the water table after the period of suction, shown by number of hours (h) in each case, is different for each stone type and is a function of small-pore mean radii.

$$dh_f = -\frac{2\mu V L}{g_c r_H^2} \tag{4.10a}$$

$$F_T = \frac{\gamma \cos \theta}{r_H g_c} \tag{4.10b}$$

where μ is the viscosity of fluid; V the mean velocity of the fluid in the pores; L the length of the capillary at height H; g_c the gravitational constant conversion factor; r_H the hydraulic mean radii of the pores, related to pore diameter as $D = 4r_H$; γ the surface tension of liquid; and θ the contact angle.

Capillaries in solid media are not cylindrical and are randomly connected. These factors are considered under the term tortuosity and expressed with the help of factor τ, defined as

$$L = \tau H \tag{4.11}$$

The mean velocity can be given as

$$V = \frac{dL}{dt} = \tau \frac{dH}{dt} \tag{4.12}$$

Substituting Eqs. (4.10) to (4.12) in Eq. (4.9), we get

$$H \frac{dH}{dt} - K_1 H - K_2 = 0 \tag{4.13}$$

where $K_1 = \rho g r^2 / 8\mu\tau^2$, $K_2 = r\gamma \cos \theta / 4\mu\tau^2$, and $r = D/2$, the mean pore radius.

At time $t = 0$, $H = 0$, and at time $t = t$, height $H = H$. Using these conditions, Eq. (4.13) can be integrated, and the final form given by

$$K_1 H - K_2 \ln \left(1 + \frac{K_1}{K_2} H \right) = K_1^2 t \tag{4.14}$$

Measuring the vertical height of water rise at different times, one can get a set of t and H values. Then Eq. (4.14) can be used to estimate mean pore diameter. Note, as Eq. (4.14) is transcendental in nature, an analytical solution is not possible but a numerical solution is. For example, for Indiana limestone with known values of different parameters, including t and H, measured experimentally,

$$\gamma = 7.28 \times 10^{-2} \frac{N}{m} \left(N = \frac{kg \cdot m}{s^2} \right), \rho = 10^3 \frac{kg}{m^3}, \mu = 3.6 \frac{kg}{m \cdot h}, \tau = 4 \text{ h}, H = 0.05 \text{ m}$$

$$K_1 = \frac{\left(10^3 \, \frac{kg}{m^3}\right)\left(9.8 \times 3600^2 \, \frac{m}{h^2}\right)[r^2(m^2)]}{(8)\left(3.6 \, \frac{kg}{m \cdot h}\right)(\tau)} = 1.1025 \times 10^9 \, r^2 \, \frac{m}{h}$$

$$K_2 = \frac{[r(m)]\left(7.28 \times 10^{-2} \times 3600^2 \, \frac{kg}{h^2}\right)}{(4)\left(3.6 \, \frac{kg}{m \cdot h}\right)(\tau^2)} = 16380r \, \frac{m^2}{h}$$

Substituting these values in Eq. (4.14) and rearranging give

$$4.0 - \frac{4.53515 \times 10^{-11}}{r^2} + \frac{1.34759 \times 10^{-14} \ln(1 + 3365.38r)}{r^3} = 0$$

Roots of this equation found by numerical methods are $r = 1.90773 \times 10^{-8}$ m = 0.019 μm. This is the mean pore radius of small pores in the Indiana limestone, which correlates well with data obtained by mercury porosimetry (Chap. 8). Mercury porosimetry provides an extremely efficient and convenient means to measure pore-size distributions in rock.

4.3.2 Salt Crystallization

When water-soluble salts are dissolved on wetting of the rock and then crystallize as the rock dries, pressure is generated in the pores. Often this pressure is large enough to cause some disintegration of the rock.

Water-soluble salts are often occluded in the pores of marine sedimentary rock (Chap. 2). They may also be drawn by capillary suction from groundwater or deposited from aerosols in seaside structures and drawn into the stone when dissolved in rainwater. Human activity is also a common source. For example, plaster and mortar used in masonry construction often contain a variety of salts. Furthermore, in cold climates, salt is often used for melting snow; disintegration of stone near street level in such regions is due to this application. Much of the destruction of limestone at the Westminster Abbey in London was due to the common salt stored in the building and that of veneer blocks in the Sphinx was due to salt-rich mortars used in various restorations.

Wetting, which is necessary to dissolve salts, occurs by a variety of ways as described before. In addition, much damage to masonry by salts occurs by the hygroscopic and deliquescent nature of these salts. Halite (NaCl), for example, will absorb moisture from humid air and go into solution. Many other sodium salts as well as salts of magnesium and potassium become liquid at a relative humidity far below that of water-vapor saturation of ambient air. Table 4.1 gives the equilibrium relative humidity and water solubility of some salt solutions.

TABLE 4.1 Equilibrium Relative Humidities at 25°C and Solubility of Some Hygroscopic Salts Occurring in Masonry

Salt	Mineral	Equilibrium Relative Humidity (%)	Solubility $(g/100 \ cm^3)$ (°C)
K_2SO_4	Arcanite	97	12 (25)
KNO_3	Niter	92.5	13.3 (0)
$Na_2SO_4 \cdot 10H_2O$	Mirabilite	87	92.7 (30)
KCl	Sylvite	84.3	34.7 (20)
Na_2SO_4	Thenardite	81	4.76 (0)
NaCl	Halite	75.3	35.7 (0)
$NaNO_3$	Soda-niter	73.9	92.1 (25)
NH_4NO_3	Nitrammite	61.8	118.3 (0)
$Mg(NO_3)_2 \cdot 6H_2O$	Nitromagnesite	52.9	125
$Ca(NO_3)_2 \cdot 4H_2O$	Nitrocalcite	50.0	660 (30)
$K_2CO_3 \cdot 2H_2O$		42.8	146.9
$MgCl_2 \cdot 6H_2O$	Bischofite	33.0	167
$CaCl_2 \cdot 6H_2O$	Antarcticite	29.0	536 (20)

Source: Modified from Arnold, A., Nature and reactions of saline minerals in walls, in *The Conservation of Stone II*, R. Rossi-Manaresi (ed.), Centro per la Conservazione delle Sculture All'Aperto, Intl. Symp. Proceedings, Bologna, 1981, p. 21.

Arnold (1982) presented a model for the salt sequence in a wall (Fig. 4.10) due to increasing dampness based upon salt solubility. In zone A, which is found at the base of the wall, the less soluble carbonates of Ca and Mg and some gypsum will precipitate. Since these salts dissolve only sparingly in water, the deterioration in this zone is less severe. In zone B, $MgSO_4$, Na_2SO_4, and KNO_3 salts are precipitated. These salts are highly soluble and have some capacity to convert moisture to liquid water. They are frequently dissolved and crystallized in varying ambient humidity. Thus, they cause severe damage to masonry. Zone C is characterized by highly hygroscopic salts of Mg, Ca, Na, and K nitrates and chlorides. Here, the salts can precipitate only if the relative humidity is extremely low. Thus Ca and Mg nitrates will almost never crystallize outside, but only in heated, very dry, interiors. Therefore, the damage to outside walls will be the least in this zone, even though the masonry looks constantly dark due to dampness. This model fits the observed conditions very well as shown by the actual distribution of salt species, the degree of disintegration, and the appearance of a dark, wet surface in zone C (Fig. 4.11).

Water-soluble salts themselves promote water transport because when in solution they exist in ionic form. Therefore, they attract water molecules and transfer them from one site to another. In this way they become an added source of stone disintegration.

It is a matter of common observation that masonry walls frequently become wet and remain so. This is due to the processes discussed before, namely:

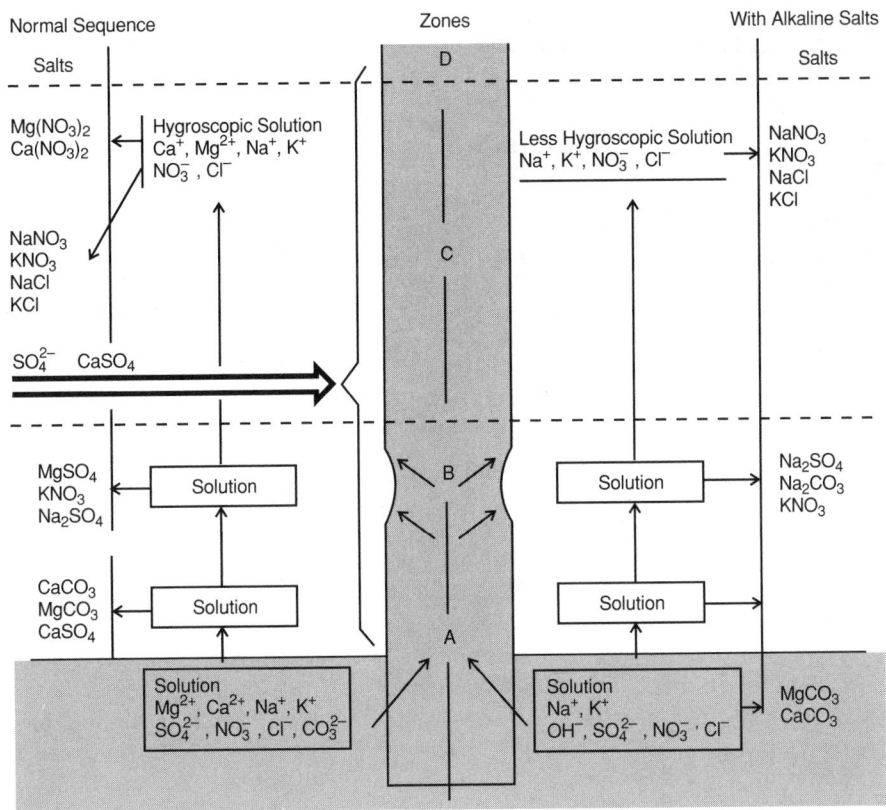

FIGURE 4.10 Model for the salt sequence in a wall. This shows that the greater the solubility of salt, the higher it will rise in a damp wall. *Source*: Arnold, A., *Rising damp and saline minerals*, in Proceedings Fourth Intl. Congress on the Deterioration and Preservation of Stone Objects, Louisville, July 7–9, 1982, Gauri, K. L. and Gwinn, J. A. (eds.), p. 21.

- Transportation of rain and underground water by capillarity
- Condensation of moisture within and outside the wall due to the fluctuation of relative humidity (see Appendix B)
- Conversion of water vapor to liquid water if hygroscopic salts are present

These walls, mostly remaining wet deep inside, become dry at and near the exterior surface by evaporation, which is controlled by temperature, relative humidity, and wind velocity. Water inside the wall then tends to move outward but it does so at a very slow rate compared with the rate of drying at surface. The reason for the slower rate of outward movement of water soaking the inside of the wall is because the mechanism of its transport, diffusion, which compared with suction due to evaporation, is an extremely slow process. As a result, a zone is created near the wall

FIGURE 4.11 Salt distribution by increasing dampness. Shown is a wall of the Franciscan Church in Solothurn, Switzerland. The light areas towards the bottom of the wall correspond to zones A and B in Fig. 4.10, and the dark upper surface corresponds to zone C in which the walls constantly remain wet due to the presence of highly hygroscopic salts. *Source*: Arnold, A., *Rising damp and saline minerals*, in Proceedings Fourth Intl. Congress on the Deterioration and Preservation of Stone Objects, Louisville, July, 1982, Gauri, K. L. and Gwinn, J. A. (eds.), p. 15.

surface, which intermittently becomes wet and dry. This zone is the site at which salts can crystallize repeatedly, thus disrupting the masonry.

A similar zone of intermittent wetting and drying is created near the surface of the wall facing the interior of building. The drying here is due to changes in relative humidity caused by the heating and cooling of the rooms, rather than by evaporation by wind, etc., at the exterior wall. Powdery deposits and flaking plaster often seen on these walls are the result of the frequent cycles of migration of the water-soluble salts leached originally from the mortar and brick from which the walls are made. If this phenomenon is confined to the lower portion of the wall, the source of salt then is mainly the subsoil.

Salts generate pressure in the pores in the following way. First, when water becomes available, the dry salt in the pore is dissolved, making a concentrated solution. When the rock begins to dry, the solution becomes saturated and crystallization of the dissolved salt begins. The growth of crystals on the pore surface exerts pressure on the pore walls, while hydrostatic pressure is developed when the crystals grow in the solution. Furthermore, certain minerals can incorporate water molecules in their crystal structure and become more disruptive due to volume increase than their anhydrous species. Anhydrous sodium sulfate (Na_2SO_4), for example, when dissolved and then crystallized as $Na_2SO_4 \cdot 10H_2O$ will generate a very high crystallization pressure. Furthermore, existing dry anhydrous salt may become hydrated in high humidity or by coming in contact with the new salt solution as it penetrates the pores. These processes of crystallization and hydration eventually result in the disassociation of grains or flaking of the rock. They are most disruptive, however, when fine capillaries are present in the rock because the capillaries are likely to become completely filled, leaving no space to accommodate the growing crystals or for the hydration pressure to be released.

4.3.2.1 Estimation of Crystallization Pressure

Pressure in Bulk Porosity. The pressure developed by the crystallization of salt can be determined quantitatively from the thermodynamic relationship of fugacity to the change in free energy of the solute:

$$\left(\frac{\partial G_T}{\partial P}\right)_T = RT\left(\frac{\partial \ln f}{\partial P}\right) = V \tag{4.15}$$

or

$$\left(\frac{\partial \ln f}{\partial P}\right)_T = \frac{V}{RT} \tag{4.16}$$

where G_T is the Gibbs free energy at constant temperature T, V the volume, P the pressure, and R universal gas constant. Also, the activity a is given by

$$\left(\frac{\partial \ln a}{\partial P}\right)_T = \left(\frac{\partial \ln f}{\partial P}\right)_T \tag{4.17}$$

If the substance is incompressible as liquid and solid, the molal volume V_s can be taken as constant, and the integration of the preceding equations give

$$\frac{V_s(P_2 - P_1)}{RT} = \ln \frac{a_2}{a_1} \tag{4.18}$$

Assuming that the solution behaves ideally, the activity a of the solute may be replaced by the concentration C of the solute.

$$P_2 - P_1 = \frac{RT}{V_s} \ln \frac{C}{C_s} \tag{4.19}$$

where P is the pressure produced by crystal growth; R is the gas constant, or 0.082 l·atm/mol·K; T is absolute temperature; V_s is the molar volume of the solid salt; C is the actual concentration of the solute during crystallization; and C_s is the concentration of the solute at saturation, that is, the solubility (g/L) of salt, for example, sodium sulfate, at given temperature.

When $C = C_s$, $P_2 - P_1 = 0$; with C exceeding C_s, crystallization pressure is produced. The effective pressure P_e generated in the sample of porosity ϕ can now be given as

$$P_e = \phi P = \phi \frac{RT}{V_s} \ln \frac{C}{C_s} \qquad (4.20)$$

Crystals will begin to grow after the solution becomes saturated with salt. Thus, only at values $C_s > C$ the pressure will be generated. The value of C/C_s can be determined experimentally as follows with respect to crystallization of sodium sulfate.

Let ϕ be the porosity of the stone and V and W be the volume and the weight of the sample, respectively. In crystallization tests (Chap. 3) before the disintegration sets in, if W_n is the weight of the sample after it has undergone n cycles, then the anhydrous salt deposited in the pores has the volume $(W_n - W)/\rho_s$, where ρ_s is the density of the salt. The new empty pore volume (V_n) after salt deposition is

$$V_n = V\phi - \frac{(W_n - W)}{\rho_s} \qquad (4.21)$$

If this sample is immersed in saturated sodium sulfate solution, the additional amount of sodium sulfate that will enter the pore space will be $V_n C_s$. Therefore, the total sodium sulfate present at this stage is

$$x = V_n C_s + W_n - W \qquad (4.22)$$

If the pores of the sample were filled with saturated Na_2SO_4 solution, the amount of Na_2SO_4 that would be present in the total pore space may be given as $V\phi C_s$. Therefore, the ratio C/C_s may be given as

$$\frac{C}{C_s} = \frac{V_n C_s + (W_n - W)}{V\phi C_s} \qquad (4.23)$$

Thus, substituting the value of C/C_s in Eq. (4.20), we have

$$P_e = \phi \frac{RT}{V_s} \ln \frac{V_n C_s + (W_n - W)}{V\phi C_s} \qquad (4.24)$$

Substituting Eq. (4.21) in Eq. (4.24) gives

$$P_e = \phi \frac{RT}{V_s} \ln \frac{[V\phi - (W_n - W)/\rho_s] C_s + (W_n - W)}{V\phi C_s} \tag{4.25}$$

We will now calculate the pressure generated in the pores of a sample of Indiana limestone after the first cycle of immersion and drying followed by immersion in the second cycle:

Volume of specimen V	24.2 ml
Porosity ϕ	11% = 0.11
Initial dry weight of sample, W	65.75 g
Weight of sample saturated with salt solution, W_n	66.60 g
$W_n - W$	0.85 g
Density of Na_2SO_4	2.68 g/ml
Solubility of Na_2SO_4 (20°C), C_s	155 g/1000 ml (anhydrous basis)
T	293 K
R	0.0821 l·atm/mol·K
Molar volume V_s	220 ml (0.22 l)

Substituting these values in Eq. (4.25),

$$P_e = (0.11) \frac{(0.0821)(293)}{0.22} \ln \frac{\left((24.2)(0.11) - \frac{0.85}{2.68}\right)(0.155) + 0.85}{(24.2)(0.11)(0.155)}$$

$$= 12.97 \text{ atm} = 190 \text{ psi}$$

This value is nearly the same as the tensile strength of Indiana limestone after the second cycle of salt crystallization (Chap. 3) when the samples begin to disintegrate.

Pressure in Small Pores. In the preceding procedure we considered the bulk porosity of the rock. Durability, however, is more precisely a function of pore-size distribution, defined as the volume of pores in pores of different radii. The following relationships can be used to calculate pressure if pore-size distributions are known (Chap. 9).

The chemical potential relates to the Gibbs free energy of a component in a system. The chemical potential of a crystal in equilibrium with a fluid subjected to a hydrostatic pressure P_1 can be expressed as

$$\mu = \mu(P_1) + v_s \gamma \frac{dA}{dV} \tag{4.26}$$

where $\mu(P_1)$ is the chemical potential of a bulk solid at pressure P_1, v_s is the molar volume of the solid, γ is the surface tension between solid and liquid, and A and V are the surface area and the volume, respectively, of the solid.

The chemical potential of a crystal growing in a small capillary, P_s, however, exceeds that of the bulk solid such that

$$\mu(P_s) - \mu(P_1) = v_s(P_s - P_1) \tag{4.27}$$

Comparing Eqs. (4.26) and (4.27),

$$P_s - P_1 = \gamma \frac{dA}{dV} \tag{4.28}$$

For a spherical interface, Eq. (4.28) can be written as

$$dP = \frac{4\gamma}{D} \tag{4.29}$$

or, for interconnected pores of variable size,

$$dP = 4\gamma \left(\frac{1}{d} - \frac{1}{D} \right) \tag{4.30}$$

where d and D are diameters of small and large pores.

Consider now the pore-size distributions in Indiana limestone in which pores of diameter around 0.1 μm cover nearly 50% of total porosity. Na_2SO_4 solution has a surface tension γ of 80 dyn/cm. Using Eq. (4.29), if all pores were of diameter $d = 0.1$ μm $= 0.1 \times 10^{-4}$ cm, then

$$dP = \frac{4 \times 80}{d} \text{ dyn/cm}^2 = \frac{(320)(1.45038 \times 10^{-5})}{d} \text{ psi}$$

$$dP = \frac{4.641 \times 10^{-3}}{0.1 \times 10^{-4}} = 464 \text{ psi}$$

Thus, the pressure generated due to the small pores (50%) would be 464×0.5 or 232 psi. The large pores in Indiana limestone have a size greater than 25 μm; therefore the pressure due to these will be very small and can be ignored in the estimation of the total pressure generated.

4.3.2.2 Estimation of Hydration Pressure
Hygroscopic salts are often present as efflorescence, that is, powdery salt encrustation due to evaporation, on and near the surface of masonry structures. Some of these salts frequently become

hydrated by incorporating molecules of water in their structure or dehydrate when these water molecules are lost, depending upon the availability of water in changing relative humidity conditions. This change in state of hydration produces large pressure, which can be estimated by the following expression:

$$P_i = \frac{nRT}{V} \ln\left(\frac{P_w}{P_h}\right) \tag{4.31}$$

where P_i is the hydration pressure; n is the number of moles of water gained by hydration; P_w and P_h are the partial pressures of water vapor and the vapor pressure of hydrated salt (mm Hg at a given temperature); $V = V_h - V_o$, where V_h and V_o are, respectively, the volume of hydrate and volume of original salt, in cm^3/mol (mol. wt./density). For Na_2SO_4 to $Na_2SO_4 \cdot 10H_2O$, the parameters are $n = 10$, $V_h = 322/1.46 = 220.54$, $V_o = 142/2.68 = 52.98$, $P_w = 17.53$, $P_h = 16.5$ (at 20°C). Substituting these values in Eq. (4.31),

$$P_i = \left(\frac{(10)(82.056)(293.15)}{220.54 - 52.98}\right) \ln\left(\frac{17.53}{16.5}\right) = 86.93 \text{ atm} = 1274 \text{ psi}$$

Pressure of this magnitude can easily cause exfoliation of masonry surfaces.

4.3.3 Frost Action

In cold climates in which the temperature fluctuates between freezing and thawing, frost is an inevitable rock breaker. The rupture of rock by freeze–thaw activity, called *frost action*, is dependant on (1) behavior of water in subfreezing temperature; (2) intensity, rate, and duration of freezing; and (3) properties of rock.

4.3.3.1 Behavior of Water in Subfreezing Temperature. In subfreezing temperature pores filled with water are likely to experience a bursting pressure, because on freezing water expands to 9% of its original volume. This process is called *frost wedging*. When the force due to expansion exceeds the tensile strength of rock or due to fatigue in cyclical freeze–thaw activity failure occurs. The pressure associated with volume increase at subfreezing temperatures is given in Table 4.2.

Bursting pressure can be generated in the frozen zone even when the rock is not fully saturated with water (Lienhart, 1993, p. 79). The main reason is that a negative pore pressure is created immediately next to the frozen front. This, combined with capillary suction, attracts water towards ice so that pores in the vicinity of the freezing front may be filled and thus experience the bursting pressure even when the pores farther away may not be completely filled and thus unaffected by frost action.

In very fine capillaries water remains unfrozen even at extremely low subfreezing temperatures due to high vapor pressure. However, this water is highly ordered and expands with decreasing temperature. This water, termed *hydration water*, is known

TABLE 4.2   **Pressure Generated by an Increase in Volume of Water on Freezing at Different Temperatures**

Temperature (°C)	Pressure (MPa)
−1.1	14
−2.8	34
−5.6	69
−9.4	103
−12.5	138
−16.7	172
−21.7	207

Source: Lienhart, D.A., The mechanism of freeze-thaw deterioration of rock in the Great Lake region, in *Rock for Erosion Control*, STP 1177, C.H. McElroy and D.A. Lienhart (eds.) ASTM, Philadelphia: ASTM, 1993, pp. 77–87.

to create high enough hydraulic pressure so that bursting will occur even though no ice has formed.

4.3.3.2   *Intensity, Rate, and Duration of Freezing.*   Freezing intensity may be defined as the critical temperature for development of strain adequate to cause crack propagation. Temperatures of −4°C to −15°C are considered to be critical for most rocks. Regarding rate of freezing, if freezing occurs slowly, ice behaves like a plastic material (Chap. 2). Therefore, a slow continued strain is developed. However, under sudden cooling, rock is more likely to rupture. Finally, the duration of freezing is also critical, given an adequate supply of moisture, because the freezing strain will continue to increase if the cycle of freezing is long.

4.3.3.3   *Rock Properties.*   Rock properties, including mechanical strength, porosity, and pore-size distributions that control water saturation, are also critical. Marble has generally a higher modulus of elasticity and practically no porosity, so it is less likely to be affected by frost action as compared with other carbonate rocks. Limestone and dolomite possess variable porosity and pore-size distributions; those with a large volume of small pores become readily and fully saturated with water and are most vulnerable to disintegration by frost action.

We will now briefly describe the implication of frost action in the quarrying and durability of rock. Deep in the earth most rocks are saturated with water. This water may be connate water, entrapped in the interstices at the time the rock was deposited; juvenile water that originated in the earth's interior; or meteoric water, derived from precipitation and filtered into the underground. Quarrying a water-saturated rock at subfreezing temperature is in any case detrimental. The immediate freezing at the surface will trap the remaining water inside the pores and soon bursting pressure will be produced after the rock is quarried. Even when the rock is quarried at temperatures above freezing, it must be dried slowly over a long period before being placed in masonry structure, a process called *curing*, because pore water is highly bonded with the capillary walls and thus hard to drive out. Furthermore, pore water contains minerals; proper curing allows for satisfactory crystallization of these minerals that then add to the strength of the rock.

4.3.4 Clay Minerals

Clay minerals often occur in carbonate rocks (Chap. 2). They may be randomly distributed like water-soluble salts or may occur concentrated in laminae in the mass of the rock. In the latter case they are highly potent agents of stone disintegration.

Clay minerals are submicroscopic particulates with grain size ranging from 0.5 μm to 4 μm. Therefore they have a high surface area to volume ratio. Two groups of clay minerals are recognized, namely, nonexpanding and expanding clays; the former have a surface area of nearly 20 m^2/g and the latter 800 m^2/g. Due to the high surface area, clay minerals have high water-adsorption capacity. On wetting they expand considerably. Furthermore, the clays have large unsatisfied charges at their surfaces. This provides them with the additional capacity to adsorb and exchange ions from circulating waters. Furthermore, when occluded in the pore space, clay minerals reduce pore size, providing the rock with an enhanced capacity for water suction. The identification of clay minerals and the quantity in which they are present are valuable aspects of the study of weathering of carbonate rocks, which is discussed in Chapter 2.

4.3.5 Thermal Expansion

Rocks, as most solids, expand when heated and contract on cooling. This may result in the disintegration of rocks.

Thermal expansion is commonly expressed as a coefficient, which is the ratio of the increase in length per unit length per °C rise of temperature. Carbonate minerals have peculiar thermal expansion characteristics. For example, the thermal expansion of calcite is 25×10^{-6}/°C along one direction (parallel to the c axis) and -5×10^{-6}/°C in a direction normal to it. Thus, the crystal expands in the c-axis direction and contracts along a- and b-axis directions when heated. However, in rocks in which the crystals are randomly oriented, the thermal expansion of crystals does not bear any relationship to the rock expansion. For example, the volumetric expansion of limestone and dolomite in the range of −20°C to 20°C is 3.4×10^{-6} and 7.7×10^{-6}/°C, respectively (Winkler, 1975, p. 44) which are rather low thermal expansion coefficients.

Observations of rocks exposed for centuries in most climatic regimes indicate that the role of thermal expansion in stone disintegration is not significant. The reasons seems to be the use of small masonry blocks in architecture and the small range of temperature experienced in most ambient conditions. However, thermal expansion may be critical when large blocks are used. To accommodate for the overall linear increase, expansion joints are provided when large blocks compose masonry structures.

SUMMARY

In natural environments in which carbon dioxide is the only chemically active gas, chemical weathering has been very slow. Therefore, monuments in nonindustrial countries or those that were removed from outdoor exposure prior to the onset of industrialization in the modern industrial world are in excellent condition.

Physical weathering, however, affects stone equally in natural and polluted environments. The agents of physical weathering are water, salts, and clay minerals. The mechanisms through which these agents affect rock are presented and mathematical expressions developed to estimate the effect of these agents.

REFERENCES

Amoroso, G. G., and Fassina V., *Stone Decay and Conservation*, New York: Elsevier, 1983.

Andrews, J. E., Brimlecomb, P., Jickells, T. D., and Liss, P. S., *An Introduction to Environmental Chemistry*, Blackwell Science, 1996.

Arnold, A., Nature and Reactions of Saline Minerals in Walls, *The Conservation of Stone II*, R. Rossi-Manaresi (ed.), Centro per la Conservazione delle Sculture All'Aperto, Intl. Symp. Proceedings, Bologna, p. 21, 1981.

Gauri, K. L. and Holdren, G. C., Jr., Pollutant Effects on Stone Monuments, *Environmental Science and Technology*, **15** (4): 386–390, 1981.

Lienhart, D. A., The Mechanism of Freeze-Thaw Deterioration of Rock in the Great Lake Region, in *Rock for Erosion Control*, STP 1177, C. H. McElroy and D. A. Lienhart (eds.), Philadelphia: ASTM, 1993, pp. 77–87.

Winkler, E. M., *Stone: Properties, Durability in Man's Environment*. 2nd ed., Wien: Springer, 1975.

CHAPTER 5

Chemical Weathering by Dry
Deposition in Polluted Environments

5.1 INTRODUCTION

Since the mid-eighteenth century when industrialization began to take place and especially since the second World War, the mechanism of decay of carbonate rock has changed and decay rates have become highly accelerated. This is due to the increased

FIGURE 5.1 Gypsum crust on marble monuments. Caryatid figures from the (left) nearly 85-year-old Field Museum of Natural History, Chicago, and (right) nearly 2,500-year-old Erechtheum on the Acropolis are shown. Both Caryatids developed gypsum crusts in the present century due to industrial sulfur dioxide emissions. The crusts on the Erechtheum Caryatid have exfoliated, whereas those on the Chicago Caryatids are yet largely intact but will fall off in the near future (see Fig. 5.3) if not treated.

use of fossil fuels: coal, and petroleum products, which contain sulfide minerals such as pyrite (FeS_2). The burning of sulfides produces sulfur dioxide (SO_2), which is reactive with carbonate minerals. Furthermore, as a result of burning at high temperature, atmospheric nitrogen is converted to nitrogen dioxide (NO_2), which is also highly reactive. As a result of SO_2 and NO_2 reactions, monuments made of carbonate rocks that survived millennia of outdoor exposure have become highly deteriorated in the present century. For example, the marble Caryatids at the nearly 80-year-old Field Museum of Natural History, Chicago, appear nearly in the same state of degradation as those at the Athenian Acropolis built nearly 2,500 years ago (Fig. 5.1).

Some believe that the quantity of SO_2 ejected in global volcanic eruptions is much larger than that ever produced by anthropogenic sources. Such eruptions have occurred throughout earth's history and yet the global atmospheric SO_2 concentration was practically negligent in the past and remains so until now in nonindustrial environments. Consequently, marble monuments currently in nonindustrial environments (Chap. 4) are in an excellent state of preservation (Fig. 5.2).

Before the influx of pollutants in industrial countries, the only chemically active gas was carbon dioxide (CO_2). However, CO_2 can only react with carbonate rocks

FIGURE 5.2 Marble monument in a nonpolluted environment. This view of Taj Mahal shows delicate carvings and carnelian (red chalcedony) inlays in the marble. Taj Mahal was built between 1635 and 1653, but the marble is in excellent condition because of the absence of sulfur dioxide in the atmosphere.

FIGURE 5.3 Effects of dry deposition and acid rain (wet deposition). Weathered marble statues, nearly 100 years old, in a Louisville, Kentucky cemetery illustrate the effects of dry deposition and acid rain. The black crust on the sheltered Madonna formed by dry deposition of sulfur dioxide has exfoliated, while the angel being washed by acid rain is highly corroded.

when it is dissolved in water. Thus, it had a minimal effect on objects protected from rain, but they are now attacked by the dry deposition of SO_2 and NO_2. Unprotected surfaces, however, then as now, dissolve in rainwater, but the rate of dissolution has increased due to the lower current pH of nearly 4 for acid rain compared with the normal rain of pH nearly 5.6. Figure 5.3 shows the effect of dry deposition and acid rain on marble statues in a Louisville, Kentucky cemetery. In this chapter we will discuss the effect of dry deposition in ambient and artificial atmospheres and describe a method to determine the source of pollutants in a region based upon the reaction product preserved on marble tombstones. In Chapter 6 we will discuss the effect of acid rain.

5.2 EFFECT OF DRY DEPOSITION OF SO₂ AND NO₂

Dry deposition consists of atmospheric gases and aerosols—tiny liquid droplets and tiny solid particles suspended in the air—being deposited on surfaces protected from the direct impact of rain. When protected from rain, the carbonate rocks react with SO_2, NO_2, and acid aerosols producing a crust, largely made of gypsum ($CaSO_4 \cdot 2H_2O$). After prolonged exposure the crusts exfoliate, causing removal of the reaction product as well as a layer of the rock a few grains thick (Fig. 5.3).

Consequently, many famous monuments over the world have been damaged beyond repair. The Caryatids of the Erechtheum at the Athenian Acropolis, mostly reduced to hunks of rock, are one such example. In the early 1980s they were removed from the Erechtheum and placed in a museum.

The exfoliation of the crust occurs by a variety of causes; most important is from a cavernous layer underneath the crust, the formation of which we will discuss later. Not only does this layer lack the original strength of the parent rock, but it is also subjected to mechanical stresses such as salt crystallization and freezing of water (Chap. 4).

5.2.1   Composition of the Crust

Prior to the mid-1980s the concentration of SO_2 greatly exceeded the concentration of NO_2. Understandably, the crusts were largely made of gypsum. Even though the modern atmosphere contains more NO_2 than SO_2, the formation of gypsum exceeds that of nitrocalcite [$Ca(NO_3)_2 \cdot 4H_2O$]. One of the reasons for this is that some NO_2 is used in oxidizing SO_2 to sulfur trioxide (SO_3), which, in turn, facilitates the sulfating reaction.

Crusts on outdoor objects lack nitrocalcite. The absence of nitrocalcite, especially on monuments of old age, can be attributed to its solubility. Nitrocalcite is a highly deliquescent salt. It absorbs moisture from air and turns it into liquid water at relative humidities as low as 50% at 25°C. Consequently, nitrocalcite drips away with the water.

Crusts formed by outdoor weathering are often black or brown in color due to soot particles embedded during growth. The soot particles are porous and contain SO_2 in their pores. In addition, many metallic particles are deposited from the environmental dust and some are contained in the rock. These particles catalyze conversion of SO_2 to SO_3, which reacts more readily with the rock, thereby enhancing the process of crust formation.

5.2.2   Mechanism of Formation and Profile of Crust

5.2.2.1   Marble.   Marble has a homogeneous composition, being almost entirely made of the mineral calcite, and is nonporous. As such, marble offers the simplest case of crust formation, the understanding of which can then be extended to other, more diverse carbonate rocks.

Scientists differ in their views on the mechanism of the formation of crust on marble. Some believe that the crust forms by the diffusion of gases into the pores of the rock. In that view, the crust grows inwards from the surface. Others, including us, believe that the crust grows outwards by the diffusion of calcium ions that then react with the gases at the surface of the growing crust. According to the latter view, the crust is an external appendage upon the original marble surface.

The mechanism of the growth of crusts in nature can be understood only when information is available for each step as the reaction proceeds. Experiments in the laboratory under controlled conditions best meet this requirement. However, the knowledge gained from laboratory experiments must be confirmed with observation made on samples that have weathered in ambient conditions for long periods of time. The following discussion is based upon studies of samples exposed in the laboratory in accelerated conditions and those that had weathered outdoors for a long period of time.

When samples are exposed in a concentrated SO_2 atmosphere at a relative humidity greater than 40% and then observed under an optical or scanning electron microscope and mineralogically analyzed by X-ray diffraction, one can see that minute stellate crystallites of calcium sulfite hemihydrate ($CaSO_3 \cdot \frac{1}{2}H_2O$) appear on the sample surface. These crystallites have distinct sites of nucleation, being separated from each other by unaltered calcite. With progression of the reaction the crystallites grow larger and eventually coalesce, forming a continuous crust of $CaSO_3 \cdot \frac{1}{2}H_2O$. However, if some moisture is able to condense, calcium sulfite crystals are oxidized and hydrated to gypsum ($CaSO_4 \cdot 2H_2O$) crystals, which have a bladed appearance. With longer exposure, the crust grows to a thickness of up to 100 μm. Figure 5.4 shows the development of crust with increasing time of exposure.

FIGURE 5.4 Formation of crust in a laboratory experiment. This scanning-electron micrograph shows formation of crust with time (h) on marble exposed in a 10 ppm SO_2 atmosphere: (A) A stellate calcium sulfite hemihydrate ($CaSO_3 \cdot \frac{1}{2}H_2O$) crystallite, 25 h; (B) coalescence of crystallites and formation of gypsum ($CaSO_4 \cdot 2H_2O$) crystals, 1,000 h; (C) entire surface covered with gypsum, 2,250 h; (D) prismatic and bladed gypsum crystals, 2,500 h. *Source*: Gauri, K. L., Kulshreshtha, N. P., Punuru, A. R. and Chowdhury, A. N., Rate of decay of marble in laboratory and outdoor exposure, *J. Mater. Civil Eng.*, **1**(2): 73–85, 1989. Reproduced by the permission of the American Society of Civil Engineers, copyright 1989 ASCE.

If these samples are cut normal to the surface, the transverse profile of the reacted and the underlying unreacted marble is revealed. This profile [Fig. 5.5(A)] shows that the reacted zone has two layers. The outer layer is the crust, made of gypsum, without the presence of any calcite. The lower surface of the crust is sharply defined and is laterally continuous, that is, it does not reveal grain boundaries, which are clearly seen in the underlying marble. The inner reacted layer is made only of calcite, is of irregular thickness, has an uneven lower surface, is more than twice as thick as the crust, and has cavities. Beneath this layer the unreacted core is present.

The purely sulfate composition of the crust with a sharp lower boundary and the porous inner reacted layer without gypsum having irregular contact with the unreacted core suggest that calcium ions must have been selectively leached from the inner reacted layer, migrating to the sample surface, and there reacting with the SO_2, resulting in the growth of crust by successive outward accretions.

A transverse profile of naturally weathered marble [Figs. 5.5(B) and 5.6] shows that the crust is made of gypsum and contains occluded soot particles that impart black color to the crust. The soot particles do not occur below the crust. However, sometimes one finds gypsum filling cavities in the underlying layer [Fig. 5.6(c)]. A comparison of this profile with that seen in controlled experiments suggests that after the crust formed, gypsum was dissolved from the crust and deposited in the

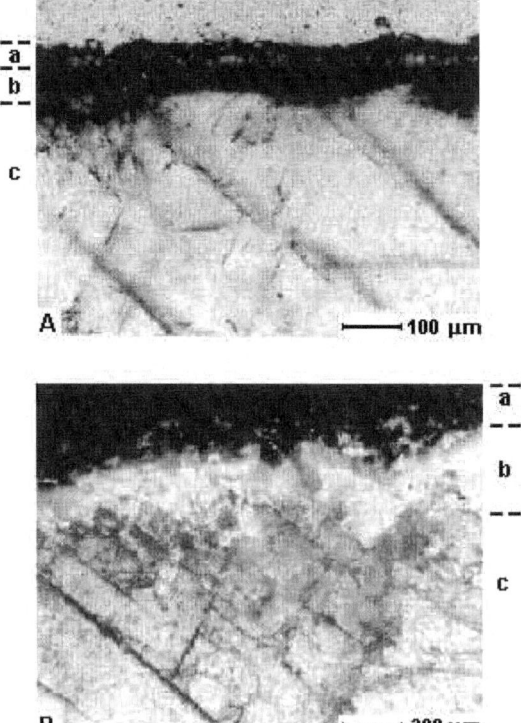

FIGURE 5.5 Profile of crust on marble reacted in the laboratory. Micrographs by light microscopy of transverse sections of Georgia marble compare: (A) marble reacted in a 10 ppm SO_2 for 3,000 h and (B) naturally weathered marble at the Field Museum of Natural History, Chicago. Both profiles show (a) the crust, (b) the leached zone, and (c) the unreacted core. *Source*: Gauri, K. L., Kulshreshtha, N. P., Punuru, A. R., and Chowdhury, A. N., Rate of decay of marble in laboratory and outdoor exposure, *J. Mater. Civil Eng.*, **1**(2): 73–85, 1989. Reproduced by the permission of the American Society of Civil Engineers, copyright 1989 ASCE.

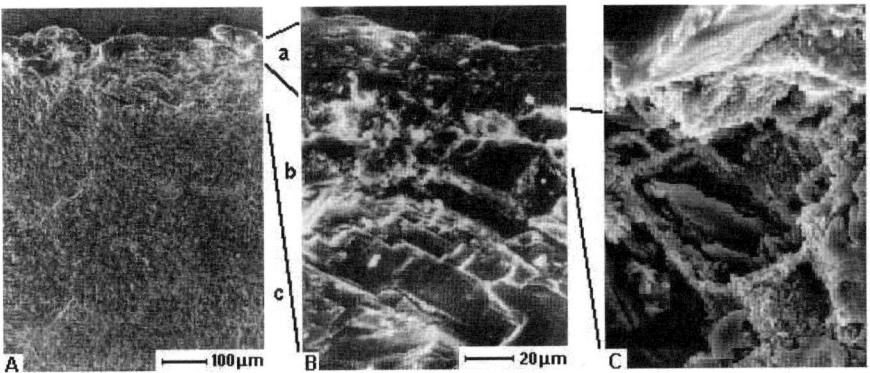

FIGURE 5.6 Profile of weathered marble. Presented are scanning electron micrographs of the transverse profile of a sample from a nearly 100-year-old marble monument in a Louisville, Kentucky, cemetery: (A) shows (a) the crust, (b) the leached zone, and (c) grains of unweathered marble; (B) crust and upper portion of leached zone; (C) upper portion of the leached zone with gypsum along cleavage planes. *Source*: Gauri, K. L., Kulshreshtha, N. P., Punuru, A. R. and Chowdhury, A. N., Rate of decay of marble in laboratory and outdoor exposure, *J. Mater. Civil Eng.*, 1(2): 73–85, 1989. Reproduced by the permission of the American Society of Civil Engineers, copyright 1989 ASCE.

inner reacted layer in prolonged outdoor exposure. Thus the mechanism of the formation of crust outdoors must be the same as that indicated by laboratory experiments.

As indicated before, in experimental conditions at room temperature and in an environment that is not saturated with water vapor (relative humidity 40% to 99%), calcium sulfite is formed and then oxidized to gypsum. The following chemical equations express these reactions:

$$CaCO_3 + SO_2 + \tfrac{1}{2}H_2O \rightarrow CaSO_3 \cdot \tfrac{1}{2}H_2O + CO_2 \qquad (5.1)$$

$$CaSO_3 \cdot \tfrac{1}{2}H_2O + \tfrac{1}{2}O_2 + 1\tfrac{1}{2}H_2O \rightarrow CaSO_4 \cdot 2H_2O \qquad (5.2)$$

Another reaction involved in the formation of crust is the catalytic oxidation of SO_2 to SO_3, as mentioned previously. The SO_3 will then react with $CaCO_3$ in the presence of water vapor to form gypsum directly without the formation of sulfite, given by

$$SO_2 + \tfrac{1}{2}O_2 \rightarrow SO_3 \text{ (g)} \qquad (5.3)$$

$$SO_2 + NO_2 \rightarrow SO_3 \text{ (g)} + NO \qquad (5.4)$$

$$CaCO_3 + SO_3 \text{ (g)} + 2H_2O \text{ (g)} \rightarrow CaSO_4 \cdot 2H_2O + CO_2 \qquad (5.5)$$

The incipient crust formed by one or several surface reactions just given secludes the underlying marble from contact with the atmosphere. However, at the surface of the crust, whenever water vapor condenses to droplets of water, the following reactions are thought to take place:

$$SO_2 + H_2O + \tfrac{1}{2}O_2 \rightarrow H_2SO_4 \tag{5.6}$$

$$H_2SO_4 \rightarrow H^+ + HSO_4^- \tag{5.7}$$

$$NO_2 + H_2O \rightarrow H^+ + HNO_3^- \tag{5.8}$$

Hydrogen ions, thus generated, penetrate through the crust and dissociate calcite in the underlying marble. The calcium ions then travel to the surface and react with the existing sulfurous and nitrous ions, resulting in the production of gypsum and nitrocalcite:

$$CaCO_3 + 2H^+ \rightarrow Ca^{2+} + H_2O + CO_2 \tag{5.9}$$

$$Ca^{2+} + HSO_4^- + 2H_2O \rightarrow H^+ + CaSO_4 \cdot 2H_2O \tag{5.10}$$

$$Ca^{2+} + 2HNO_3 + 4H_2O \rightarrow 2H^+ + Ca(NO_3)_2 \cdot 4H_2O \tag{5.11}$$

Skoulikidis and Charalambous (1981) consider the crust as analogous to patina on metals. After a thin crust of gypsum is formed by surface reactions, they explain its growth on the basis of a solid-state diffusion and galvanic cell model, expressed by the electrochemical reactions

$$SO_2 + \tfrac{1}{2}O_2 \rightarrow SO_3 \quad \text{(catalyzed reaction)} \tag{5.12}$$

$$(+)SO_3 (g) + \tfrac{1}{2}O_2 (g) + 2e^- \rightarrow SO_4^{2-} (s) \tag{5.13}$$

$$(-)CaCO_3 (s) \rightarrow Ca^{2+} (s) + CO_3^{2-} (aq)$$
$$\rightarrow Ca^{2+} (s) + CO_2 (g) + \tfrac{1}{2}O_2 (g) + 2e^- \tag{5.14}$$

$$CaCO_3 + SO_3 \rightarrow CaSO_4 \cdot 2H_2O (s) + CO_2 (g) \tag{5.15}$$

Here, SO_3 is considered as the positive pole, $CaCO_3$ of the virgin marble as the negative pole, and the early formed crust as the electrolyte.

5.2.2.2 Dolostone.
Dolostone is made of the mineral dolomite [$CaMg(CO_3)_2$]. The dolomite–SO_2 reaction can be given as

$$CaMg(CO_3)_2 + 8H_2O + 2SO_2 + O_2$$

$$\rightarrow CaSO_4 \cdot 2H_2O + MgSO_4 \cdot 7H_2O + 2CO_2 \tag{5.17}$$

The crust on dolostone, in natural weathering, is made of gypsum only [Fig. 5.7(A)]. Under experimental conditions, however, the formation of epsomite, $MgSO_4 \cdot 7H_2O$, is found and can be ascertained as follows.

After a short period of exposure in laboratory experiments at relative humidity below the dew point, one often finds drops of water on the stone surface. When some drops of liquid are drawn from the sample surface, slowly dried, and X-ray diffraction traces taken, the dried material turns out to be largely epsomite, $MgSO_4 \cdot 7H_2O$. This suggests that epsomite must have formed in outdoor weathering but has not been preserved. The reason for this is that epsomite is a deliquescent salt. It converts moisture at 88.3% relative humidity (20°C) to liquid water, and like nitrocalcite in the case of marble drips away.

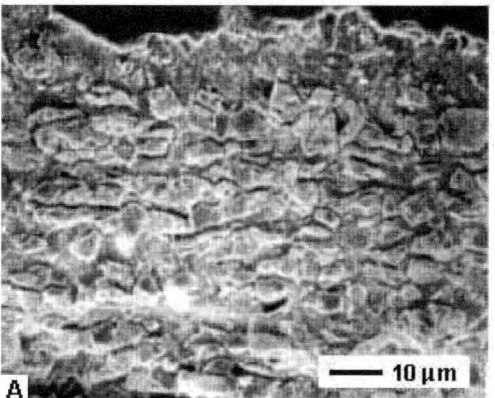

FIGURE 5.7 Profile and surface view of weathered dolostone. Presented are scanning electron micrographs of Laurel dolomite samples from the façade of the nearly 120-year-old St. Louis Bertrand Church in Louisville, Kentucky. (A) Transverse profile showing the gypsum crust above and the unaltered dolomite crystals with typical rhombohedral outline. (B) Surface view showing open pores. *Source*: Gauri, K. L., Tambe, S. S., and Caner-Saltik, E. N., Weathering of Dolomite in Industrial Environments. *Environ. Geol. Water Sci.*, **19**(1): 55–63, 1992. Reproduced by the permission of Springer-Verlag, copyright 1992 Springer-Verlag.

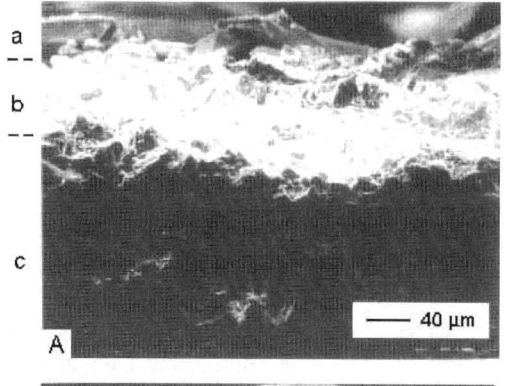

a
--
b
--

c

A

—— 40 μm

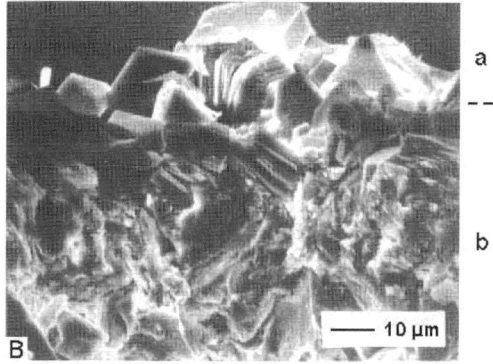

a
--

b

—— 10 μm

B

FIGURE 5.8 Profile of dolostone reacted in a laboratory. Presented are scanning electron micrographs of the transverse profile of a Laurel dolomite sample reacted for 410 h in a 10 ppm SO_2 atmosphere, showing (a) the gypsum crust, (b) the leached zone, and (c) unreacted dolostone. *Source*: Gauri, K. L., Tambe, S. S. and Caner-Saltik, E. N., Weathering of Dolomite in Industrial Environments. *Environ. Geol. Water Sci.*, **19**(1): 55–63, 1992. Reproduced by the permission of Springer-Verlag, copyright 1992 Springer-Verlag.

The profile of an outdoor-weathered dolostone, a porous rock, is surprisingly similar to that of practically nonporous marble. One would expect that due to the presence of pores, SO_2 would diffuse into the pore space and react with surrounding grains, leaving the cores of the grains unaffected. Instead, the crust is entirely made of gypsum [Fig. 5.7(A)] and no dolomite is found in the crust. However, some differences between the crust on dolomite and that on marble are apparent.

First, in the weathered zone of the dolostone the inner leached layer, corresponding to the inner layer clearly demarcated in marble, is not well developed [Fig. 5.7(A)]. The reason is that the original large pores in the dolostone cannot be distinguished from space newly created by leaching.

Second, dolostone has open pores in and at the surface of the crust [Fig. 5.7(B)], which seem to be absent in the marble. This can be easily explained by the fact that pores in dolostone are large and act as conduits for the migration of calcium and magnesium ions. They are like volcanic vents, which continue to grow outward as new material is precipitated around them. Thus the size and shape of the original pores in the dolostone are maintained in the crust. The pores are also probably maintained in the case of marble; however, the intergranular space in marble is extremely narrow and the closely packed cleavage planes from which the ions move are just too narrow in the crust to be clearly visible.

The transverse profile of the sample of dolostone exposed in a laboratory experiment in a sulfur dioxide–enriched atmosphere is similar (Fig. 5.8) to that of dolostone exposed outdoors. When the surface of such a laboratory-reacted sample is observed under a microscope, gypsum crystals are seen [Fig. 5.9(A)]. At a higher magnification, the bladed morphology of gypsum crystals enclosing the pores becomes evident [Figs. 5.9(B) and (C)].

These observations lead to the conclusion that a common mechanism is responsible for the formation of crust in natural weathering and in laboratory conditions both for nonporous marble and porous dolomite. Since limestone ranges in porosity between marble and dolomite, the same mechanism can be considered to be operative in varieties of limestone, which we have confirmed by many laboratory and field studies.

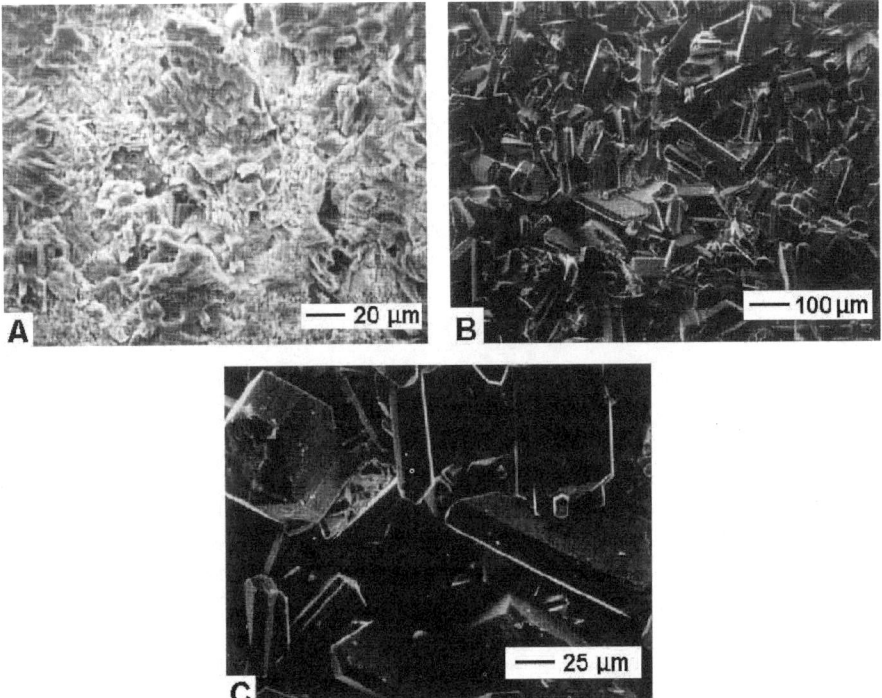

FIGURE 5.9 Surface of dolostone reacted in a laboratory. Presented are scanning electron micrographs of surfaces of Laurel dolomite samples reacted in a 10 ppm SO₂ atmosphere. (A) Reacted for 120 h showing scattered gypsum crystals; (B) reacted for 410 h showing the entire surface covered with large gypsum crystals; (C) a portion of (B) enlarged, showing an open pore in the middle of the figure. *Source*: Gauri, K. L., Tambe, S. S. and Caner-Saltik, E. N., Weathering of Dolomite in Industrial Environments. *Environ. Geol. Water Sci.*, **19**(1): 55–63, 1992. Reproduced by the permission of Springer-Verlag, copyright 1992 Springer-Verlag.

5.2.3 Laboratory Setup to Simulate the Formation of Crust

As stated earlier the mechanism and the rate of growth of crust can be best determined through experiments in controlled laboratory conditions. We now describe the experimental setup in which these reactions can be carried out.

Slabs are cut from the rock. Since most calcareous rocks inherently contain gypsum, this gypsum must be removed before the samples are exposed to SO_2 atmosphere. This can be done by a continuous washing of samples with tap water for weeks followed by washing with deionized water until the salt is completely removed, which can be determined by analysis for sulfates by ion chromatography. The samples are then polished with silicon carbide powder to produce smooth and uniform surfaces, cleaned ultrasonically in deionized water to remove abrasive and calcite dust, dried at 105°C, cooled in a desiccator to room temperature, and then weighed for density determination and measured for surface area. The last two measurements are needed to estimate crust thickness. Whereas the linear measurements of length and width of the sample provide the surface area, the density can be measured by the use of a pycnometer. (See Appendix C.)

A pycnometer is a device used to measure the volume of a sample accurately. It is a cylinder that can be fully filled with water without any trapped air, which is achieved by the lid of the cylinder being concave and having a hole in the center. Four weight measurements are needed to determine the density: (1) the weight of the water-saturated sample, W_1; (2) the weight of the water-filled pycnometer, W_2; (3) the weight of pycnometer with the sample and filled with water, W_3; and (4) the weight of the dry sample (W'). The bulk volume (V) of the sample and its density (D) are given by

$$V = \frac{W_1 + W_2 - W_3}{d} \qquad (5.16)$$

$$D = \frac{W'}{V} \qquad (5.17)$$

where d is the density of water, which is 1. Density (D) can be easily calculated by substituting corresponding values in the program provided at the Web site.

Dry samples are exposed in a reactor (Fig. 5.10) that can be made by modifying a desiccator so that it has an inlet and an outlet for entry and exit of the artificial atmosphere, which is created by mixing SO_2 and NO_2 from a concentrated source with a carrier gas. A constant flow rate of the carrier gas can be obtained by using electronic mass-flow controllers. A precise amount of concentrated SO_2 and NO_2 is obtained from commercially available permeation tubes, devices which deliver a constant quantity of the gas at a given temperature. For example, to produce a 10 ppm SO_2 atmosphere, a flow (800 cm^3/min) of clean air is passed over the tube with a permeation rate of 1690 ng/min for SO_2 at 20°C. The permeation rate can be checked by weight loss of the permeation tube over time and confirmed by measurements of these gases at the exit of the reactor by gas analyzers. Alternatively, the SO_2-enriched atmosphere may be passed through deionized water to which hydrogen peroxide (H_2O_2) has been added. H_2O_2 oxidizes SO_2 to H_2SO_4. This solution can then be analyzed by ion chromatography for $(SO_4)^{2-}$ ions and converted to SO_2 by multiplying by the factor $\frac{8}{9}(M_{SO_2}/M_{SO_4})$, where M is the molecular or ionic weight.

Dry SO_2 does not react with calcite. Humidity can be produced and adjusted by placing water in the reactor and by passing the carrier air through a water bubbler. Also, salts of variable hygroscopicity can be used to produce the desired humidity. To produce high humidity, essential in carrying out experiments at a reasonably quick rate, the SO_2–air stream is injected in the reaction chamber beneath the water table; the samples are placed in the reactor only after the water has been equilibrated with the given concentration of SO_2. For NO_2 atmosphere, the gas must be introduced above the water column because this gas is highly reactive with water.

To ensure that the reaction proceeds with a desired dose of reactive gas, the quantity of the gas at the reactor entrance and exit should nearly be the same, which requires that a small number of samples be exposed simultaneously.

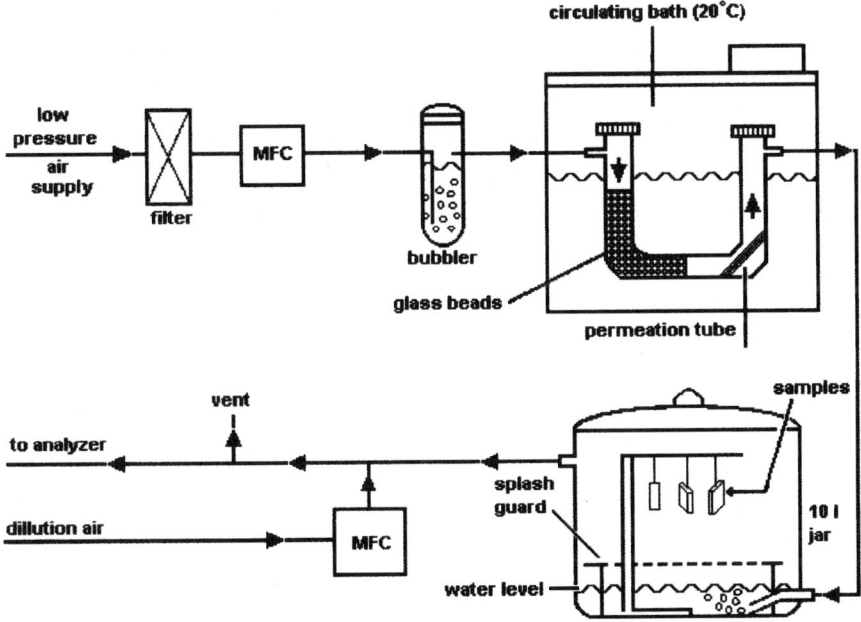

FIGURE 5.10   Reaction chamber. This is the experimental setup for reactions in concentrated atmospheres. MFC is an electronic mass-flow controller to provide for a stable and accurate flow of air that is then mixed with a known dose of SO₂ or NO₂ to obtain the desired atmosphere. SO₂ or NO₂ can be obtained from permeation tubes that, at a constant temperature, release a constant volume of the respective gas. The permeation tubes are placed in baths and the entire setup is also lodged in a constant-temperature room to maintain a reasonably constant condition for the kinetic study.

To estimate the rate of reaction we need to measure the crust thickness over time. The following gives some methods through which this can be done.

5.2.4 Measuring Crust Thickness

The crust thickness can be measured by many techniques, including (1) microscopic observations using light and scanning-electron microscopy (SEM) of sectioned surfaces; (2) mapping for elementary sulfur (S) in the transverse profile by energy dispersive X-ray (EDS) techniques used in conjunction with SEM; (3) successive abrasion of the crust until the entire thickness of the crust has been removed as determined by (a) X-ray diffraction (XRD) identification of the mineralogical species and (b) X-ray photoelectron spectroscopy (XPS) measurement for elementary S and N; and (4) dissolving the reaction product in deionized water and ion chromatography (IC) of the solution. These methods are concisely described in the following.

5.2.4.1 *Microscopic and EDS Measurements* Microscopic and EDS measurements require preparation of transverse sections by cutting the samples normal to the outer surface and grinding the cut surface to smoothness. This often requires embedding the samples in polymers so that the crust does not fall off in the process of grinding. The polished surfaces can then be viewed under an optical or electron microscope and the crust thickness measured. SEM is the method of choice because of its high resolution and infinite depth of focus, whereby even an uneven surface is projected in a planar definition. If the SEM is fitted with an EDS system, then the measurement of the distance from the surface to the depth at which sulfur is present gives the crust thickness (Fig. 5.11).

These methods are useful when a large crust thickness is present, as on samples that have weathered outdoors for long periods or samples that have been reacted in the laboratory in a high concentration of active gases for many hours.

5.2.4.2 *Measurement by XRD* To obtain crust thickness by XRD, thin layers of crust are abraded in some lapping device fitted with a micrometer and X-ray diffractograms (Chap. 2) taken of the freshly exposed surfaces. This operation is continued until gypsum disappears (Fig. 5.12).

5.2.4.3 *Measurement by XPS* Photoelectron spectroscopy, like EDS, allows for quantitative measurement of elements present at a surface. When fitted with a plasma source for sputter-depth profiling, XPS offers an elegant means to determine the crust thickness. Figure 5.13 shows the surface of samples exposed for 3 and 6 months in a Louisville, Kentucky environment with prevailing concentrations of 10 ppb and 30 ppb SO_2 and NO_2, respectively. The exposed surface was analyzed for N, S, and Ca and then abraded by helium plasma so that 1 minute of sputtering denuded an 18-Å-thick layer. Successive abrading and XPS measurements until no more S and N appeared marked the lower boundary of the crust. It can be seen that a crust nearly 0.3 μm and 0.6 μm thick had formed on Georgia marble in 3 and 6 months, respectively.

This method is appropriate for samples that have been exposed outdoors for short periods and thus have a thin crust. For thicker crusts, this method is not practical because of the expensive instrument time needed to perform this measurement.

5.2.4.4 Ion Chromatography
Ion chromatography measures precisely, at levels of parts per billion (ppb), the quantity of cations or anions in solution. The exposed sample is leached in deionized water and the solution analyzed for sulfate and nitrate anions (Fig. 5.14). Depending upon the thickness of the crust, sometime weeks are needed to leach the entire reaction product.

The quantity of sulfate and nitrate ions can be converted, by mass balance, to the respective quantities of gypsum and nitrocalcite. To obtain the respective volumes, the mineral masses are divided by their densities, 2.32 and 1.89, respectively. The crust thickness δ is then calculated by the following relationship:

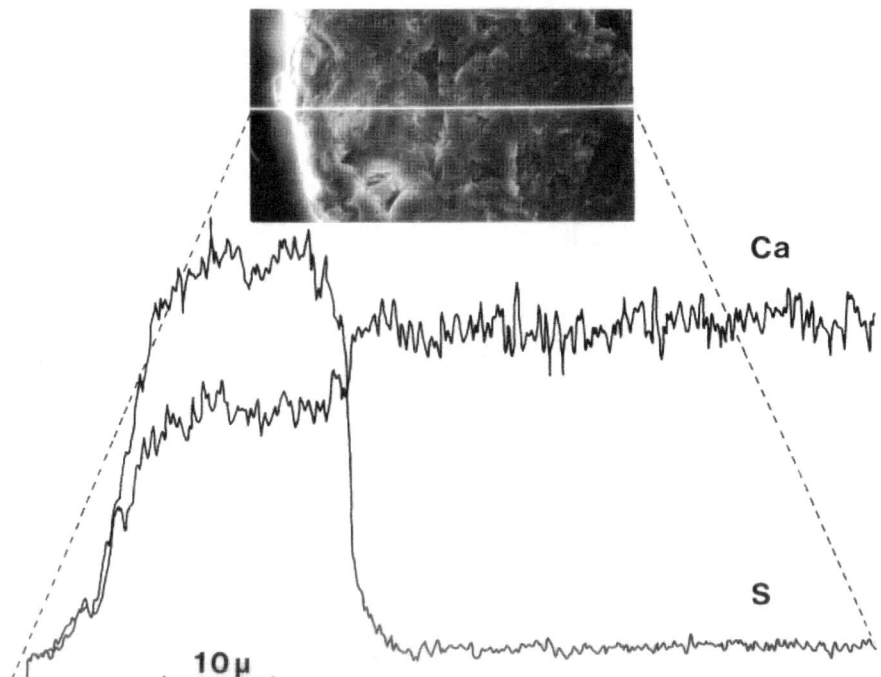

FIGURE 5.11 Measurement of crust thickness by SEM and EDS. A scanning electron micrograph (above) and energy-dispersive X-ray (EDS) tracing of sulfur (S) and calcium (Ca) shows the presence of gypsum crust in a sample of Carrara marble reacted with a 10 ppm SO₂ atmosphere for 1,450 h. Note the absence of S below the crust. The scale at the bottom refers to the EDS trace. *Source*: Gauri, K. L., Kulshreshtha, N. P., Punuru, A. R. and Chowdhury, A. N., Rate of decay of marble in laboratory and outdoor exposure, *J. Mater. Civil Eng.*, **1**(2): 73–85, 1989. Reproduced by the permission of the American Society of Civil Engineers, copyright 1989 ASCE.

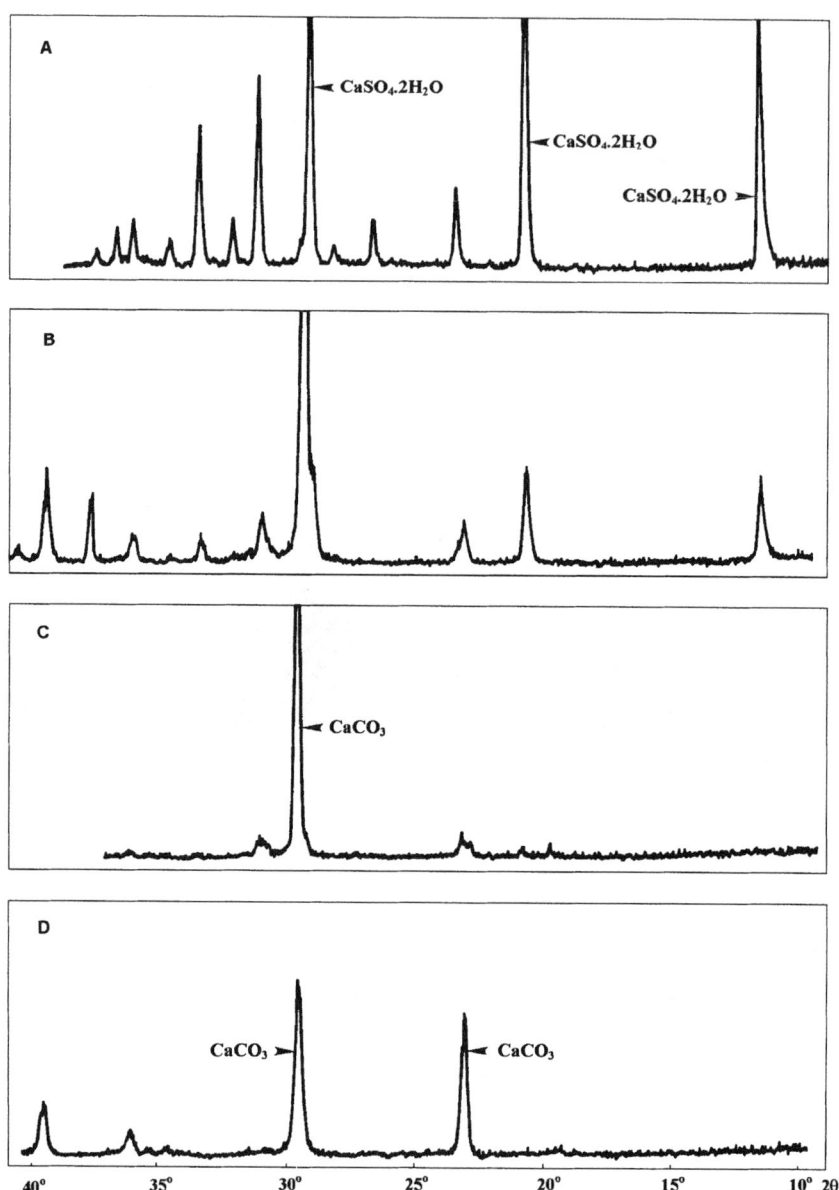

FIGURE 5.12 Measurement of crust thickness by X-ray diffraction. X-ray diffractometer tracings of successively abraded surfaces of Georgia marble weathered beneath a cornice at the Field Museum of Natural History, Chicago, show (A) gypsum at the external surface of the black crust; (B) gypsum at a brown surface within the crust; (C) calcite at a surface within the leached zone, nearly 120 μm from external surface; (D) a surface in the unweathered portion of marble. *Source*: Gauri, K. L., Effect of acid rain on structures, in *Acid Rain*, Proceedings, ASCE, National Convention, Gunnerson and Willard (eds.), 1975, p. 75. Reproduced by the permission of the American Society of Civil Engineers, copyright 1979 ASCE.

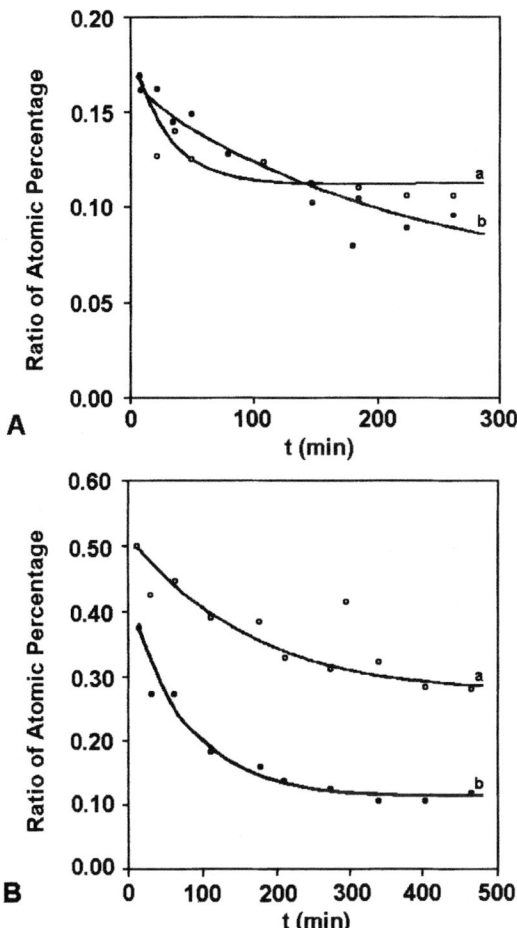

FIGURE 5.13   Measurement of crust thickness by XPS. X-ray photoelectron spectroscopy (XPS) of surfaces produced from sputter-depth profiling by helium plasma of samples of Georgia marble exposed under a shed in a Louisville environment for (A) 3 and (B) 6 months. The specimen surfaces were abraded with a helium plasma so that an 18-Å-thick layer was removed in 1 minute. Curve (a) shows the ratio of Ca to S and curve (b) the Ca to N ratio. It is estimated that 150 min and 300 min were needed to remove the crust, giving thicknesses of 0.27 μm and 0.54 μm formed in the 3 and 6 month periods, respectively. *Source*: Yerrapragada, S. S., Jaynes, J. H., Chirra, S. R. and Gauri, K. L., Rate of weathering of marble due to dry deposition of ambient sulfur and nitrogen dioxides. *Anal. Chem.*, **66**(5): 655–659, 1994. Reproduced with permission of the American Chemical Society, copyright 1994 ACS.

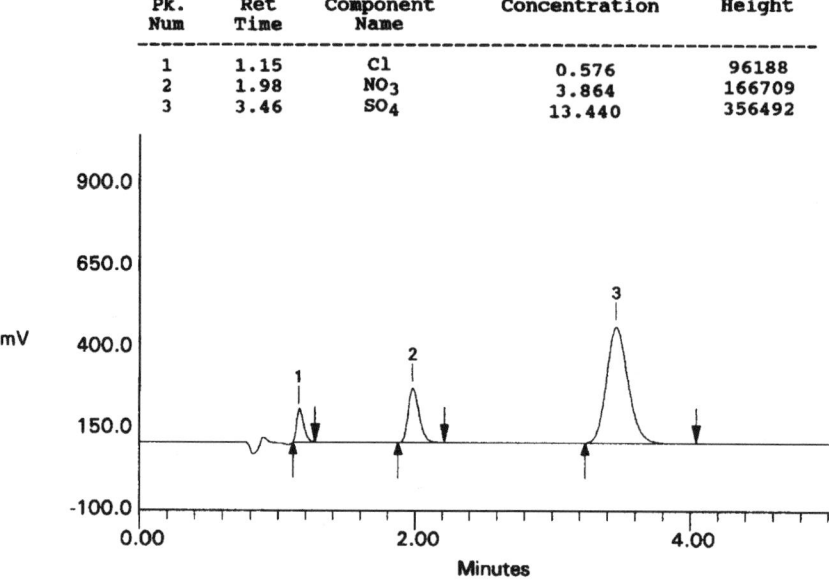

Pk. Num	Ret Time	Component Name	Concentration	Height
1	1.15	Cl	0.576	96188
2	1.98	NO$_3$	3.864	166709
3	3.46	SO$_4$	13.440	356492

FIGURE 5.14 Measurement of crust thickness by ion chromatography. The concentrations (ppm) of sulfate and nitrate ions are shown, which can be converted to crust thickness. The sulfate and nitrate ions were extracted by immersing the sample in deionized water.

$$\delta = \left(\frac{1.8 m_{SO_4}}{2.32} + \frac{3.8 m_{NO_3}}{1.89} \right) / A_0 \qquad (5.18)$$

where m_{SO_4} and m_{NO_3} are the masses of sulfate and nitrate ions and A_0 is the surface area of the sample before exposure.

5.3 REACTION RATES

This section deals with the rates of reaction determined from data collected from laboratory experiments. In the next chapter we will show that these rates can be used to predict the magnitude of chemical weathering in any outdoor condition.

5.3.1 Marble

Marble is made primarily of the mineral calcite. Therefore, its reaction with SO_2 can be described by Eqs. (5.5) and (5.10). Figure 5.15 describes a general SO_2–marble reaction in which the concentration of the reaction product with time is plotted. It can be seen from the trend of the rate curve that in early stages the reaction is fast but slows down after some reaction product has accumulated. This suggests that the initial reaction rate is kinetically controlled. In later phases, however, a thick product layer

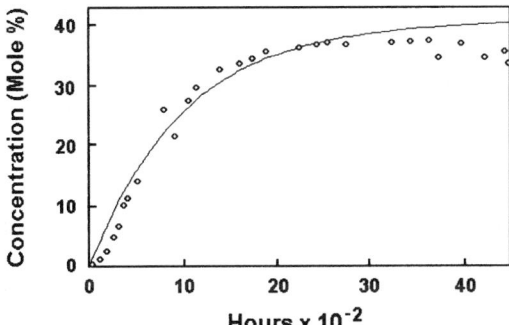

FIGURE 5.15 Trend of the reaction-rate curve of marble. The concentration of the reaction product with time is shown. Notice that initially the reaction is fast but slows down as the crust thickens. The reaction in the initial stage is considered kinetically controlled and in later stage, diffusion controlled.

offers resistance to outward migration of Ca^{2+} ions; thus the rate becomes diffusion controlled, which is a slow transport mechanism.

Marble is a compact rock. Therefore, porosity is not a factor in determining the reaction rate. However, in different varieties of marble the grain size varies. Accordingly, reaction rates vary from one variety of marble to another. Other factors that would control the rate of reaction are the gas concentration and relative humidity. We will now describe the effects of these variable factors on the reaction rates.

5.3.1.1 Effect of Grain Size Figure 5.16 shows the rate curves for two varieties of marble, namely, the Georgia and Alabama marbles with average particle diameters of 1,000 μm and 150 μm, respectively, and for a briquette prepared by pressing crushed marble particles with a diameter less than 38 μm. The briquette [Fig. 5.16(a)] shows a maximum reaction because of the large surface area of the small particles. Accordingly, Alabama marble [Fig. 5.16(b)] is attacked more than Georgia marble [Fig. 5.16(c)].

5.3.1.2 Effect of Gas Concentration The effect of gas concentration is evident from the rate curves shown in Fig. 5.17. It is apparent that the rate of the reaction is nearly proportional to the gas concentration. We will show later that these curves can be expressed by rate constants, which are valid for SO_2 concentrations in the range prevailing in ambient atmospheres.

5.3.1.3 Effect of Relative Humidity Relative humidity plays a major role in the rate of the reaction as shown by Fig. 5.18. Dry SO_2 does not react with calcite. At high humidity, approaching 100% relative humidity, the reaction product formed is nearly twice as much as that at 60% to 80% relative humidity. The reaction product at up to 80% relative humidity is all calcium sulfite hemihydrate [Eq. (5.1)]. Above 80% relative humidity, some calcium sulfite changes to gypsum [Eq. (5.2)]. At 100%

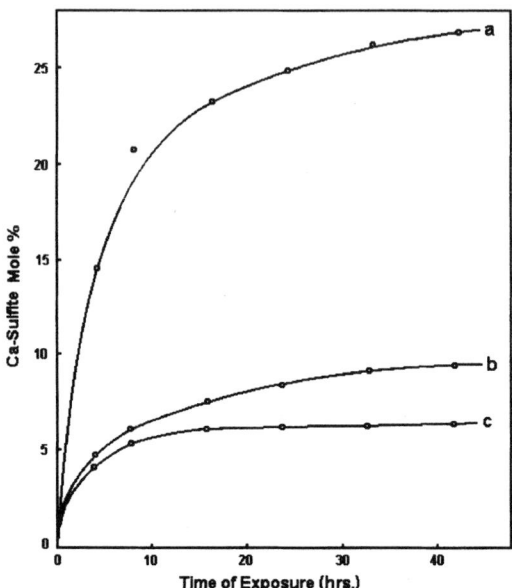

FIGURE 5.16 Effect of particle size (μm) on the reaction rate of marble. Reaction-rate curves for marble types of variable grain size are plotted: (a) powdered marble, <38 μm; (b) Alabama marble, 150 μm; (c) Georgia marble, 1,000 μm, average particle diameter. Notice that the fine-particulate marble shows the maximum and the coarse-particulate marble the least reaction. See Chapter 1 for SEM micrographs of marble types. *Source*: Gauri, K. L., Popli, R. and Sarma, A. C., Effect of relative humidity and grain size on the reaction rates of marble at high concentration of SO_2, *Durability Building Mater.*, **1**: 209–216, 1982/83.

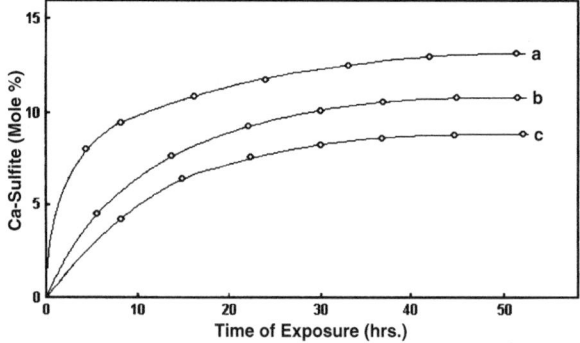

FIGURE 5.17 Effect of SO_2 concentration on the reaction rate of marble. Rate curves for SO_2 concentrations: (a) 3,000 ppm, (b) 1,500 ppm, (c) 700 ppm, showing that the reaction rate is nearly proportional to the gas concentration. *Source*: Gauri, K. L., Popli, R. and Sarma, A. C., Effect of relative humidity and grain size on the reaction rates of marble at high concentration of SO_2, *Durability Building Mater.*, **1**: 209–216, 1982/83.

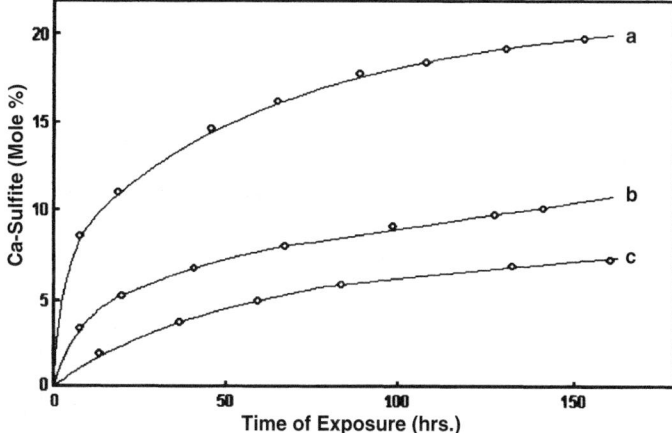

FIGURE 5.18 Effect of humidity on the reaction rate of marble. The reaction-rate curves in variable relative humidity are plotted: (a) 95%–100%, (b) 75%–80%, (c) 45%–50%. *Source*: Gauri, K. L., Popli, R. and Sarma, A. C., Effect of relative humidity and grain size on the reaction rates of marble at high concentration of SO_2, *Durability Building Mater.*, **1**: 209–216, 1982/83.

relative humidity, if water vapor is occasionally allowed to condense as droplets on the stone surface, all calcium sulfite changes to gypsum.

5.3.2 Limestone

Common limestone varieties are porous. Therefore, unlike marble, the rate of reaction in limestone is controlled by pore-size distributions. We will describe reactions for four limestone varieties, namely, Cordova Cream (CR), Cordova Shell (CS), Louder

TABLE 5.1 Porosity and Pore-Size Distributions in Some Limestone Varieties

Sample[a]	Pore Size			Porosity (%)
	>50 μm	10–50 μm	<10 μm	
CR	0.0536[b]	0.0024[b]	0.0830[b]	26
CS	0.0570[b]	0.0210[b]	0.0580[b]	22
LD	0.0520[b]	0.0040[b]	0.0370[b]	20
BD	0.0300[b]	0.0050[b]	0.0420[b]	14

[a]The samples are from the limestones (Chapter 1): CR = Cordova Cream, CS = Cordova Shell, LD = Louder, and BD = Bedford or Indiana limestone.
[b]The pore volume, shown for each category of pores, is in cm^3/g.

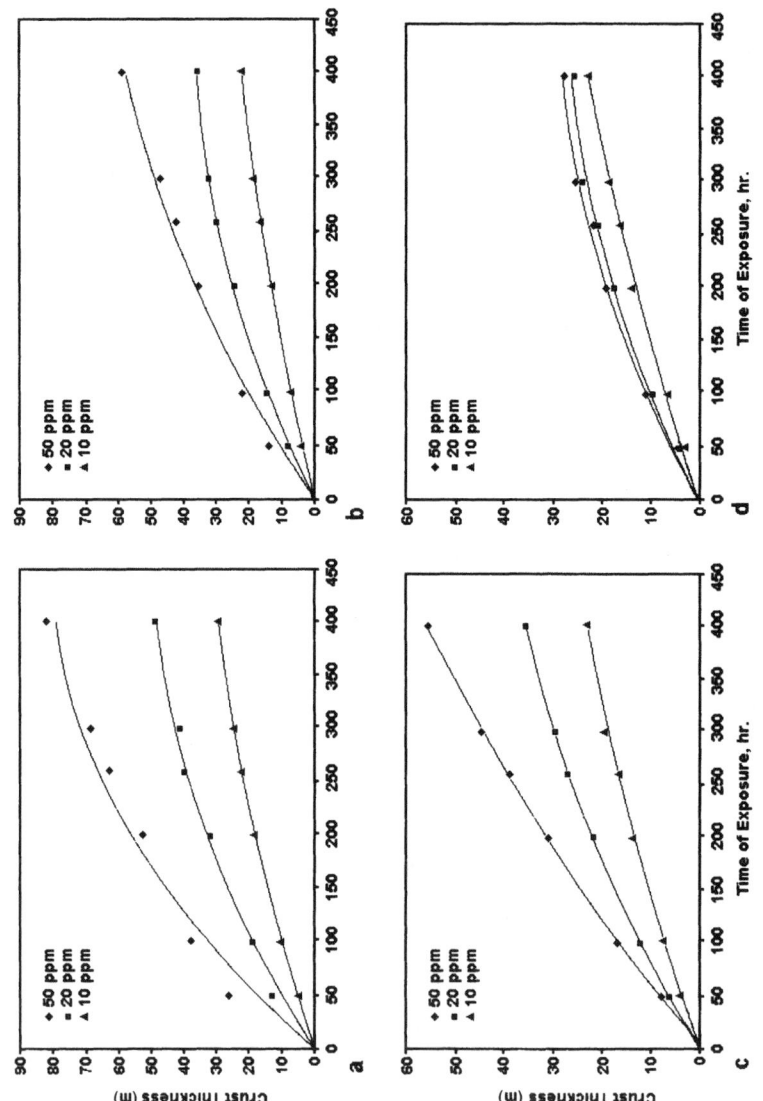

FIGURE 5.19 Effect of pore size on the reaction rate of limestone. Rate curves in variable SO₂ concentration for limestone are plotted: (a) Cordova Cream, (b) Louder, (c) Cordova Shell, and (d) Bedford limestone. The curves are drawn on the basis of regression analysis. *Source:* Bandyopadhyaya, J. K., Annamalai, S. and Gauri, K. L., Application of artificial neural networks in modeling limestone-SO2 reaction, *Am. Inst. Chem. Eng. J.*, **42**(8): 2295–2301, 1996. Reprinted with permission of the American Institute of Chemical Engineers, copyright 1996 CHE.

(LD), and Bedford (BD) described in Chapter 1, the pore-size distributions of which vary considerably (Table 5.1).

The profiles of the limestone conversions in 10 ppm, 20 ppm, and 50 ppm SO_2 atmospheres are shown in Fig. 5.19. CR shows the maximum conversion and BD the least because CR has the largest porosity as well as largest volume of small pores that promote reaction by offering large surface area. Furthermore, the curves show declining reaction rates particularly with prolonged exposures, somewhat similar to that observed with marble.

5.3.3 Dolostone

Figure 5.20 shows the reaction of Laurel dolomite. The reactions were carried out in atmospheres of nearly 10 ppm, 15 ppm, and 25 ppm SO_2. However, knowing that the magnitude of reaction is nearly proportional to the concentration, a unified rate curve plotted as a product of concentration and time versus the reaction product (crust thickness) can represent the entire data. We notice that the rate curve shows, unlike in marble and limestone, an acceleration of the reaction rate with time. The reasons, though not fully known, appear to be related to the porosity of the stone and the

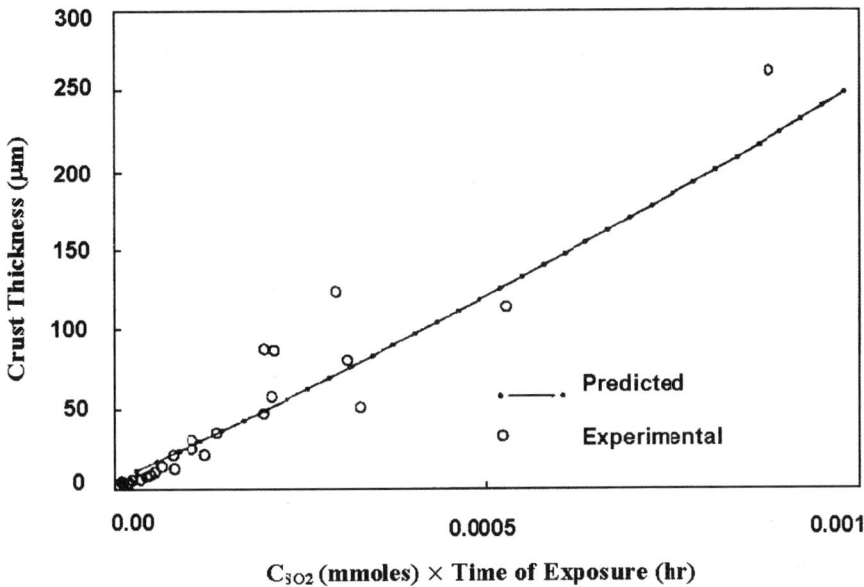

FIGURE 5.20   Reaction-rate curve for Laurel dolomite. The reaction was carried out in a nearly 10 ppm, 15 ppm, and 25 ppm SO_2 atmosphere. The curve is a quadratic polynomial fit for the data points obtained as a product of concentration and time versus crust thickness. Notice a slight increase in the reaction rate with time possibly related to the formation of epsomite, a deliquescent salt. *Source*: Gauri, K. L., Tambe, S. S. and Caner-Saltik, E. N., Weathering of Dolomite in Industrial Environments. *Environ. Geol. Water Sci.*, **19**(1): 55–63, 1992. Reproduced by permission of Springer-Verlag. Copyright 1992 Springer-Verlag.

wetness of surface caused by the deliquescent reaction product epsomite ($MgSO_4 \cdot 7H_2O$).

5.4   ACIDITY GRADIENT

The rate of change of SO_2 and NO_2 concentrations in a region, or the acidity gradient, largely depends upon the source of their emission. Concentrations of these gases are commonly measured using sophisticated gas analyzers that are costly and quite expensive to maintain. We will now describe a novel inexpensive technique using lithomonitors through which acidity gradients can be determined and thus the sources of pollutants identified.

A lithomonitor is defined as a rock object that can act as a fingerprint of the acidity. Requirements of a lithomonitor are that it be made of a homogenous rock and be easily reacted with atmospheric acids, and that dated objects of this rock be found frequently over a large area. Marble tombstones possess these requirements. Alternatively, chips of marble can be exposed outdoors over time. In the following we give results of a study (1987–88) in the KIPDA region (an acronym for Kentucky–Indiana Planning and Development Agency) in the United States, which includes the metropolis of Louisville and several rural counties surrounding it.

Scrapings were obtained from tombstones of Georgia marble from several cemeteries. The selected tombstones were either located under arches or, when exposed, had deeply carved lettering in which the reaction product was preserved. These scrapings were analyzed for their sulfate content. The dates of installation of monuments were marked on a map where the sample analysis in the laboratory had revealed the occurrence of sulfate. Figure 5.21 shows contours that represent the time of the appearance of sulfate. For example, the area enclosed by the contour 1975 is the region in which sulfate had formed on *certain* monuments installed since 1975. In the area outside this contour, all monuments with sulfate were older. The shorter period of exposure in which sulfate appeared indicates a relatively higher rate of acid deposition.

A decreasing gradient in acid deposition in the KIPDA region is discernable in all directions from the Ohio Valley area of Jefferson County. The region of higher acid deposition nearly coincides with the location of power plants.

These gradients do not express the actual concentration of the gases as the term gradient in the normal sense would imply. Nevertheless, they give empirically the trends of the distribution of the actual concentrations.

This approach to pinpoint the source of SO_2 emissions can have a useful application in resolving the problem of potential weathering of carbonate monuments, such as Taj Mahal. Numerous studies have been made to predict SO_2 levels in the environs of Taj Mahal resulting from emissions of a refinery nearly 30 km away. Another purported source of emissions are the foundries in Agra—the city where Taj Mahal is located—threatened for closure by the government of India. The case is being litigated in the Supreme Court of India as to the real source of emissions. A study of the type just outlined can easily solve this problem while also showing

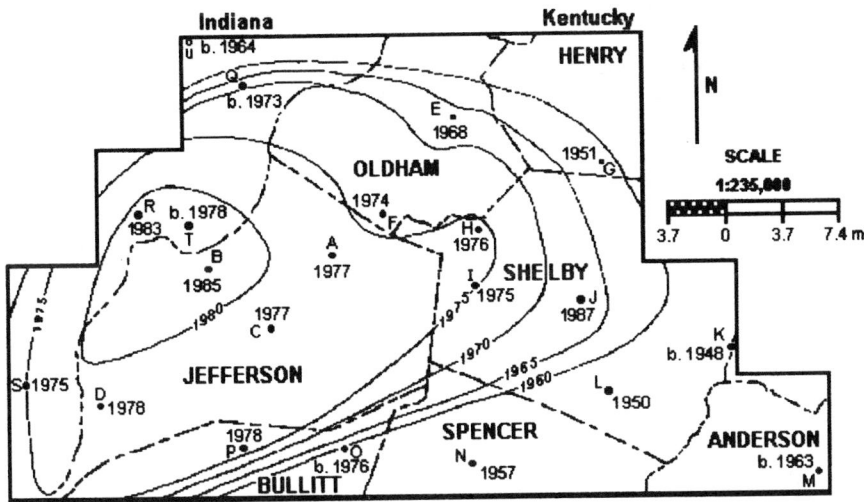

FIGURE 5.21 Acidity map of the KIPDA region showing time contours. Each point represents the year of installation of the monument with sulfate. The samples were collected in 1987. *Source*: Gauri, K. L., Punuru, A. R. and Holdren, G. C., Acidity gradients in the KIPDA region, *Environ. Geol. Water Sci.*, **15**(1): 55–58, 1990. Reproduced by the permission of Springer-Verlag, copyright 1990 Springer-Verlag.

whether the marble of Taj Mahal is in imminent danger of the decay that has visited all marble monuments in the industrialized world.

SUMMARY

In an atmosphere containing oxides of sulfur and nitrogen, monuments made of carbonate rocks develop crust largely made of gypsum. The crusts eventually exfoliate, often within 100 years of installation, destroying the sculptural form.

The mechanism of crust formation in nonporous marble and porous limestone and dolomite is the same. After the initial surface reaction the crust grows outward by the reaction of the calcium and magnesium ions, leached from unaltered rock beneath the incipient crust, with atmospheric gases. This mechanism is clarified by the study of rocks weathered outdoors and samples exposed in accelerated laboratory conditions.

The reaction rates of marble, limestone, and dolomite in sulfur dioxide concentrations ranging from 10 ppm to 50 ppm are given. It is suggested that rate constants can be developed to predict the rate of weathering in ambient gas concentrations, discussed in Chapter 6.

Lastly, a method based on the study of dated monuments is described whereby the semiquantitative acidity gradient in a region can be determined. This method also allows identification of the source of pollution.

REFERENCE

Skoulikidis, Th., and Charalambous, D., Mechanism of Sulfation of atmospheric SO_2 of the limestone and marbles of ancient monuments and statues, *Brit. Corrosion J.*, **16** (70): 1981.

CHAPTER 6

Kinetics and Modeling Decay Rates of Carbonate Rocks in Polluted Environments

6.1 INTRODUCTION

In Chapters 4 and 5 we dealt with the decay mechanisms of carbonate rocks in natural and polluted environments. In Chapter 5 we also gave reaction rates of rock samples exposed in concentrated atmospheres of sulfur dioxide simulating the dry-deposition effect of polluted environments. Now, we will use those data in mathematical models to develop rate parameters, such as the order of the reaction, the rate constant, the rate of diffusion, and so on, which can then be used to predict the reactivity of carbonate rocks in any ambient polluted environment. The models used here are commonly applied in the field of chemical engineering for gas–solid noncatalytic reactions, but we have found them useful in modified forms to evaluate air-pollutant–carbonate reactions.

Furthermore, as indicated in Chapter 4, carbonate-rock surfaces exposed to rain are subject to dissolution and are thus eroded, or reduced, in time. The degree of surface reduction is largely a function of the volume and the pH of rain. Based on data obtained by exposing samples to the ambient environment in the Louisville, Kentucky area, we will also develop relations in this chapter that can allow prediction of surface reduction with any known pH level and amount of rainfall.

6.2 MODELING THE GROWTH OF CRUST

Carbonate rocks protected from rain in outdoor weathering have crusts that are mainly composed of gypsum ($CaSO_4 \cdot 2H_2O$) as explained in Chapter 5. Such crusts form as a result of dry deposition of sulfur dioxide. However, controlled outdoor experiments also show the formation of calcium nitrate that forms due to the nitrogen dioxide reaction but is not preserved due to its high solubility. We will first discuss the growth of crust by sulfur dioxide reactions and later include in the model the reaction due to nitrogen dioxide.

The rate of the growth of crust can be modeled on the basis of the production of gypsum, the conversion of calcite, or the amount of sulfur dioxide deposited. The reaction can also be modeled for the change in porosity. These are the quantities that we have determined through experiments, some of which have been described in Chapter 5. We will use these properties in a shrinking unreacted core model, a deposition velocity model, a percolation theory, and a random pore model. We will also use artificial neural networks (ANN) to model these reactions. The development of these models, except ANN, is based upon the common fundamental controls of gas–solid reactions, namely:

- *External Mass Transfer.* In order to react with the sample, the gaseous reactant must reach the sample surface. This involves the mass transfer of the reactant from the airstream through the boundary layer surrounding the particle. Adjacent to any solid body there is a thin layer of air held stationary by friction. This layer, called the boundary layer, offers resistance to the penetration of the reactant gas. This resistance is termed *aerodynamic resistance*.

- *Surface Chemical Reaction.* This takes place between the reactive gas and the solid following kinetics, the parameters of which include the acid-buffering capacity of carbonate rocks, temperature, gas concentration, stoichiometry, and order of reaction. The resistance offered by the object to react is termed *surface-uptake resistance*.

- *Internal Diffusion.* For the reaction to continue after the surface reaction has taken place and a thin crust formed, the reactant gas must either penetrate through the crust or the constituents of rock travel outwards to come into contact with the reactant gas at the sample surface. This transport, largely affected by the product-layer resistance, is considered under internal diffusion. In the following models the internal diffusion relates to the outward migration of calcium ions as we have shown in Chapter 5. Generally, however, internal diffusion is considered as the diffusion of the gaseous reactant through the reaction product. Nevertheless, in both cases the situation is analogous in that the reaction front comes into being, which migrates inward in time shrinking the inner unreacted core.

In the mathematical derivations that follow the mass transfer coefficient (h_d) and the kinetic rate constant (k_s) are expressed in cm/h and the internal diffusion (D_e) in cm^2/h.

6.3 SHRINKING UNREACTED CORE MODEL

The shrinking unreacted core model, or simple shrinking-core model, has been frequently used to describe gas–solid reactions (Szekley et al., 1976, Mazet and Spinner, 1992). The shrinking-core model is graphically represented by Figure 6.1, which shows the boundary layer (a), where C_{Ab} and C_{As} are the concentrations of the gas outside the layer (bulk concentration) and inside it at the sample surface (surface

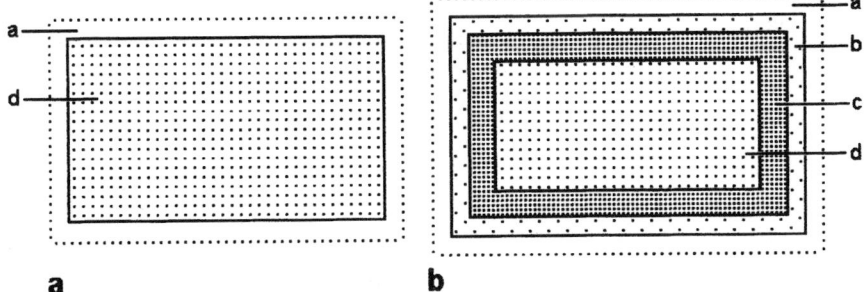

FIGURE 6.1 Dry-deposition reaction. Schematic diagrams for (a) unreacted and (b) reacted carbonate rock specimens. (b) shows the gas film (a), crust (b), leached zone (c), and unreacted core (d). The reactant gas diffuses through the gas film to reach the sample surface, where the surface reaction results in the formation of thin crust. Calcium ions then migrate from the leached zone to the surface, whereby the crust grows in thickness. As the reaction proceeds, the reaction front between the leached zone and the unreacted core advanced inwards.

concentration). The crust (b) is shown exterior to the original sample surface; the outer layer of the original sample is the leached zone (c) and its inner surface is the reaction front. Underneath the leached zone is the unreacted core (d).

The rate-controlling parameters, the mass-transfer coefficient (h_d), the kinetic rate constant (k_s), and the internal diffusion (D_e) can be treated as inverses of corresponding resistances. The overall reaction rate then can be given as the sum of the resistances, as

$$r_A = \left(\frac{1}{R_1 + R_2 + R_3} \right) A_0 C_{As} \tag{6.1}$$

or

$$r_A = \left(\frac{1}{h_d} + \frac{1}{k_s} + \frac{\delta}{D_e} \right)^{-1} A_0 C_{As} \tag{6.2}$$

where r_A (mol/h) is the rate of reaction, R_1 is the aerodynamic resistance, R_2 is the surface-uptake resistance, and R_3 is the product-layer resistance, expressed by symbols shown in Eq. (6.2); δ (μm) is the product-layer thickness; and A_0 (cm^2) and C_{Ab} (mol/cm^3) are the external surface area of the sample and the concentration of the gaseous reactant in the bulk, that is, in the airstream, above the boundary layer.

The net rate, r_A, is with respect to the rate of consumption of the gas, which we will consider later in the development of the sulfur dioxide deposition model. This rate can be related to the solid reactant conversion (r_B).Consider the general reaction

$$aA(g) + bB(s) \rightarrow cP(g) + dQ(s) \tag{6.3}$$

where A and B represent reactants, P and Q the reaction products, and a, b, c, and d the stoichiometric coefficients.

Thus, the consumption of A at rate r_A and that of B at r_B can be expressed as

$$\frac{r_A}{a} = \frac{r_B}{b} \tag{6.4}$$

and the rate of reaction with respect to the conversion of solid B can be given as

$$r_B = \frac{\rho_B}{M_B} \frac{dV_R}{dt} \tag{6.5}$$

where, ρ_B and M_B denote the density (g/cm^3) and molecular weight, respectively, and V_R is the volume of the reacted rock in time t (h). For the reaction

$$CaCO_3 + SO_2 + 2H_2O + \tfrac{1}{2}O_2 \rightarrow CaSO_4 \cdot 2H_2O + CO_2 \tag{6.6}$$

The mass-transfer-controlling, reaction-controlling, and the diffusion-controlling rates are given by

$$r_m = A_0 h_d (C_{Ab} - C_{As}) \tag{6.7a}$$

$$r_k = A_0 k_s C_{As}^n \tag{6.7b}$$

$$r_D = \frac{A_0 D_e C_{As}^n}{\delta} \tag{6.7c}$$

where, δ is the crust thickness, C_{As} and C_{Ab} are the concentration of SO_2 (mol/cm^3) at the sample surface and outside the boundary layer, and n is the order of reaction, found to be a first order reaction. C_{As}, is obtained from C_{Ab} by knowing the mass-transfer coefficient of SO_2 through the boundary layer, the calculation of which we will show later. When all the resistances are present, the net rate (r_A) of a gaseous reactant is given by Eq. (6.2).

Using Eqs. (6.4) and (6.5), the conversion of the gaseous reactant can express the conversion of solid reactant as

$$\frac{\rho_B}{M_B} \frac{dV_R}{dt} = \frac{b}{a} \left(\frac{1}{h_d} + \frac{1}{k_s} + \frac{\delta}{D_e} \right)^{-1} A_0 C_{As} \tag{6.8}$$

which after substitution and rearrangement becomes

$$\frac{d\delta}{dt} = \alpha \frac{M_B}{\rho_B} \left(\frac{1}{h_d} + \frac{1}{k_s} + \frac{\delta}{D_e} \right)^{-1} C_{As} \tag{6.9}$$

and has the analytical solution

$$\left(\frac{1}{2D_e} \right) \delta^2 + \left(\frac{1}{h_d} + \frac{1}{k_s} \right) \delta = \alpha \frac{M_B}{\rho_B} C_{As} t \tag{6.10}$$

where α is the stoichiometric coefficient, being 1 for reaction represented by Eq. (6.6); δ is the crust thickness for C_{As}, the concentration of SO_2 (mol/cm^3) at the sample surface in time t; ρ_B and M_B are the density (g/cm^3) and molecular weight, respectively, of the reacted rock; h_d is the mass transfer coefficient; D_e is the internal diffusion or internal effective diffusivity, and k_s is the kinetic rate constant.

The data presented in Chapter 5 show the crust thickness (δ) over time in known gas concentrations in the bulk (C_{Ab}). We will show how the mass-transfer coefficient (h_d) and the reactant concentration at the sample surface (C_{As}) can be calculated from the atmospheric flow-rate data.

6.3.1 Mass-Transfer Coefficient and Surface Concentration

The value of the mass-transfer coefficient, h_d, is needed primarily for evaluation of laboratory experiments in which it is often difficult to produce fast airstreams; thus their flow rate is in the laminar range. The air turbulence outdoors, however, frequently mixes the air mass so that the same concentration can be postulated to occur at the sample surface as it is in the bulk air.

The mass-transfer coefficient is a function of several aerodynamic properties of the fluid, such as viscosity, density, and velocity, and can be given by

$$h_d = \frac{D_{AB} N_{Sh}}{d_p} \tag{6.11a}$$

where D_{AB} is the binary diffusion coefficient of the air–SO_2 system, N_{Sh} is the Sherwood number, and d_p is the equivalent diameter of the experimental slab. The values of D_{AB} (cm^2/s) are calculated using the Chapman–Enskog equation:

$$D_{AB} = 0.0018583 \frac{\sqrt{T^3 (1/M_A + 1/M_B)}}{p \sigma_{AB} \Omega_{AB}} \tag{6.11b}$$

where T is temperature in K, M_A and M_B are the molecular weights of air and SO_2, respectively, p is the pressure in atmosphere, and σ_{AB} and Ω_{AB} are known from literature (Bird et al., 1960).

N_{Sh} can be obtained from the Rantz–Marshall correlation for the Reynolds number (N_{Re}) in the range of laminar flow as

$$N_{Sh} = 2.0 + 0.6 N_{Re}^{0.5} N_{Sh}^{0.33} \tag{6.12}$$

where N_{Re} and N_{Sc} denote the Reynolds and Schmidt numbers. The Reynolds number is proportional to the ratio of inertial to viscous effects and the Schmidt number is the ratio of molecular momentum diffusivity to molecular mass diffusivity. Their values are obtained using the definitions

$$N_{Re} = \frac{d_p V \rho_g}{\mu}, \quad N_{Sc} = \frac{\mu}{\rho_g D_{AB}} \tag{6.13}$$

The notations V, μ and ρ_g represent, respectively, the linear fluid velocity (cm/s), viscosity, and density of humid air. The value of μ is found from viscosity tables as 0.001827 (g·cm^{-1}·s^{-1}) at 293 K.

Now we will calculate the values of D_{Ab} following Eq. (6.11b). For the system air (A)–$SO_2(B)$, $\sigma_A = 3.617$ and $\sigma_B = 4.29$. Thus, $\sigma_{AB} = (\sigma_A + \sigma_B)/2 = 3.9549$; Ω_{AB} is a dimensionless function of the temperature and of the intermolecular potential field for one molecule of A and one molecule of B, at 25°C the value of $\Omega_{AB} = 1.084$; $M_A = 28.97; M_B = 64.064; \mu = 0.000183$ g/cm·s; $\rho_g = 0.001205$ g/cm^3.
Therefore,

$$D_{AB} = 0.0018583 \frac{\sqrt{298.15^3(1/28.97 + 1/64.064)}}{(1)(3.9549^2)(1.084)} = 0.12644 \frac{cm^2}{s}$$

Next, the N_{Re} and N_{Sc} values needed for N_{Sh} can be obtained with the following parameters: Jar diameter $d_j = 12.5$ cm; particle diameter $d_p = 2.07$ cm; flow rate = 300 cm^3/min; and the fluid velocity is

$$V = \frac{\pi d_j^2}{4 \times (\text{flow rate})} = \frac{\pi \times 12.5^2}{4 \times 300 \times 60} = 0.040727 \frac{cm}{s}$$

Substituting these values in Eq. (6.13),

$$N_{Re} = \frac{2.07 \times 0.040727 \times 0.001205}{0.000183} = 0.555$$

$$N_{Sc} = \frac{0.000183}{0.001205 \times 0.12644} = 1.1993$$

Thus,

$$N_{Sh} = 2.0 + (0.6 \times 0.555^{0.5} \times 1.1993^{0.33}) = 2.47506$$

TABLE 6.1 Mass-Transfer Coefficients

Flow Rate (cm³/min)	Mass-Transfer Coefficient (cm/h)
300	544.27
400	560.43
500	574.67
600	587.54
700	599.37
800	610.39
900	620.74
1000	630.52

Note: Experimental flow conditions are jar diameter, 12.5 cm; equivalent particle (sample) diameter, 2.07 cm; $T = 298.15$ K.

Finally, the mass-transfer coefficient is

$$h_d = \frac{0.12644 \times 2.4750}{2.07} = 0.1511 \ \frac{\text{cm}}{\text{s}} = 544.2694 \ \frac{\text{cm}}{\text{h}}$$

Table 6.1 shows the mass-transfer coefficient calculated for typical flow rates used in laboratory experiments. We also give programs at the ftp site (see Appendix C) through which the mass-transfer coefficient for any set of parameters can be calculated. Once the mass transfer coefficient is obtained, the surface concentration can be calculated using Eq. (6.7a).

6.3.2 Rate Constants with Respect to SO_2 Reactions

Now, we know from experiments the SO_2 concentration at the sample surface and the thickness of the crust at any time during the experiment. Hence, the surface rate constant (k_s) and the internal diffusion (D_e) can be found as unique values by the regression analysis of the experimental data. These constants for various varieties of marble, limestone, and dolomite are given in Table 6.2. Figures 6.2 through 6.4 show the experimental data and the rate curves drawn on the basis of the model equation [Eq. (6.10)].

6.3.3 Rate Constant with Respect to the $SO_2 + NO_2$ Reaction

We have noted previously that laboratory experiments for SO_2 in controlled and accelerated conditions provide the best means to model the reactions mathematically. However, the reactions in the presence of NO_2 are difficult to perform because of the reaction of NO_2 with SO_2 and the complex NO_2–water reaction. In a series of

TABLE 6.2 **Rate Constants and Effective Diffusitvities for Marble, Limestone, and Dolomite Reactions in Sulfur Dioxide Atmosphere**

Sample Type	Name	Surface Rate Constants k_s (cm/h)	Effective Diffusivity D_e (cm^2/h)
Marble[a]	Carrara marble	312	0.14
Limestone[a]	Cordova Cream	509.80	0.55
	Cordova Shell	288.88	0.41
	Louder	224.13	0.45
	Bedford	134.02[b]	0.083[b]
Dolomite[a]	Laurel dolomite	183.10	0.37

[a]For description of the types of marble, limestone, and dolomite, see Chapter 1.
[b]The order of reaction for all rocks is 1 except for Bedford limestone where for reaction in 10 ppm and 20 ppm sulfur dioxide atmosphere, the order is 0.96.

experiments that we performed, the reaction occurred only in one experimental set up shown in Figure 6.5. Even in this case the reaction stopped after nearly 200 h with SO_2 and NO_2 concentrations of 10 ppm each.

Shortly after the reaction started, droplets of water began to form on the sample surface. Analysis by X-ray diffraction of the solid remaining after evaporating some of these drops on a glass slide at nearly 40°C revealed nitrocalcite [$Ca(NO_3)_2 \cdot 4H_2O$] while the dried sample surface showed gypsum ($CaSO_4 \cdot 2H_2O$) only. Nitrocalcite is a deliquescent salt that can condense moisture from the airstream. Through careful

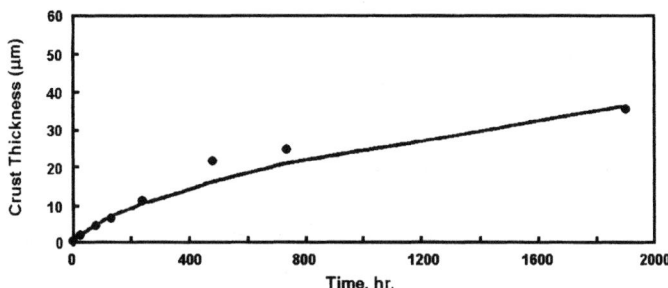

FIGURE 6.2 Marble reaction in 10 ppm SO_2–atmosphere: application of the shrinking-core model. The curve shows the model-predicted crust thickness after Eq. (6.10) and the dots represent actual data points. The fitting of the model equation give values for k_s and D_e as 312 cm/h and 0.14 cm^2/h. *Source*: Yerrapragada S. S., Chirra, S. R., Jaynes, J. H., Bandyopadhyay, J. K. and Gauri, K. L., Weathering rates of marble in laboratory and outdoor conditions, *J. Environ. Eng.*, **122**(9): 856–863, 1996. Reproduced with the permission of the American Society of Civil Engineers, copyright 1996 ASCE.

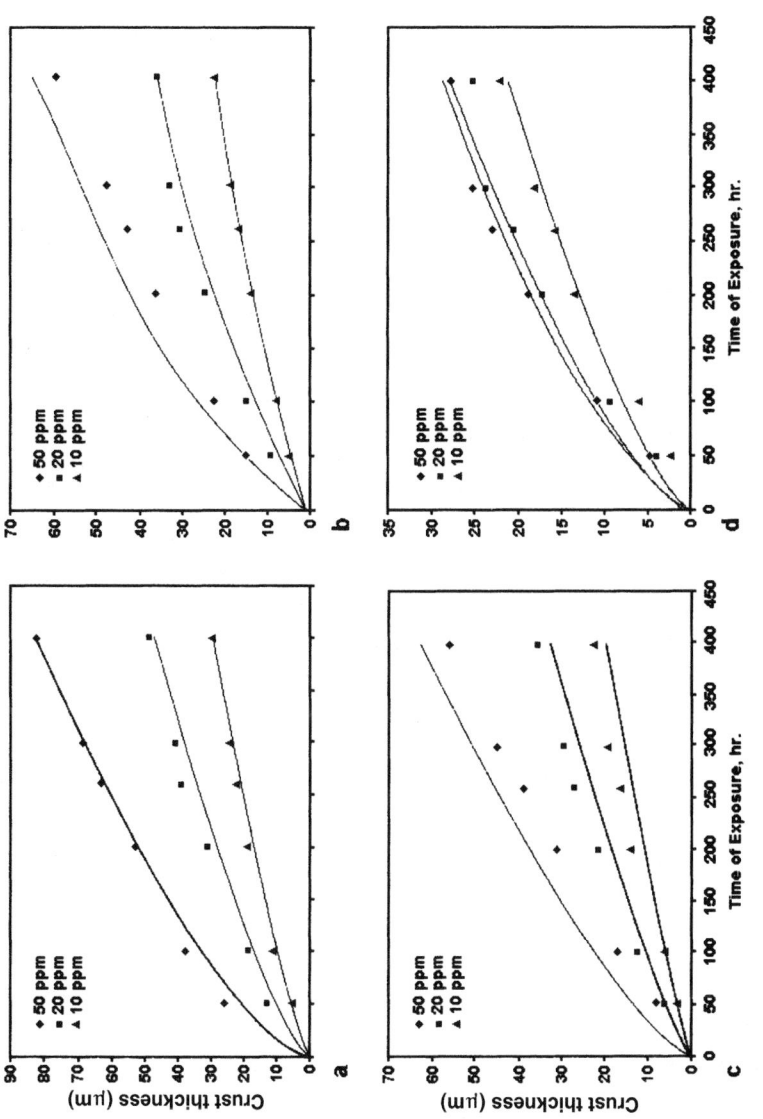

FIGURE 6.3 Limestone reaction in variable SO₂-concentrations: application of the shrinking-core model. The curves show the model-predicted crust thickness after Eq. (6.10) and dots represent actual data points. Limestone types are (a) Cordova Cream; (b) Cordova Shell; (c) Lueders; and (d) Bedford. *Source:* Bandyopadhyay, J. K., Annamalai, S., and Gauri, K. L., Application of artificial neural networks in modeling limestone–SO2 Reaction, *Am. Inst. Chem. Eng. J.,* **42**(8): 2295–2302, 1996. Reproduced with the permission of the American Institute of Chemical Engineers, copyright 1996 AIChE.

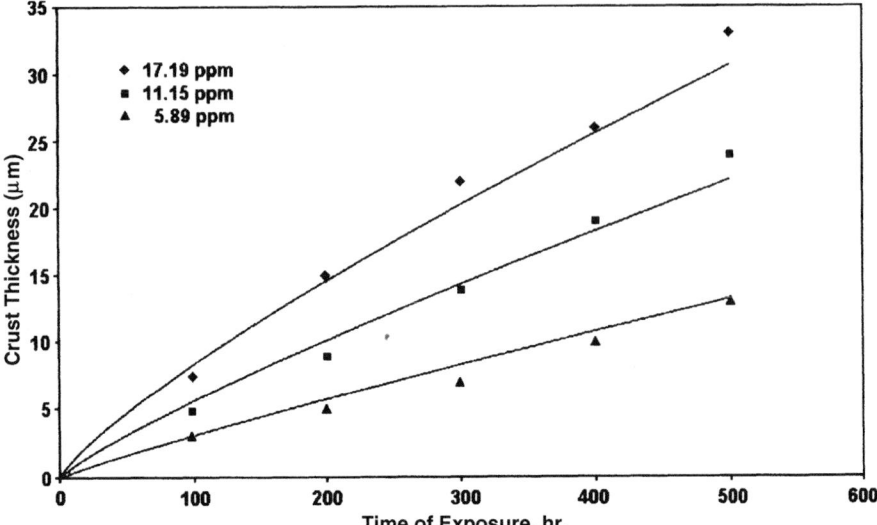

FIGURE 6.4 Dolomite reaction in variable SO_2 concentrations: Application of the shrinking-core model. The curves show the model-predicted crust thickness after Eq. (6.10) and dots represent actual data points. This is for Laurel dolomite.

handling of these specimens to avoid dripping away of nitrocalcite, analysis by ion chromatography for sulfate and nitrate ions showed their ratio was 0.7:0.3. These findings allow modeling of the $SO_2 + NO_2$ reaction.

The formation of crust due to the $SO_2 + NO_2$ reaction, based on XRD analysis, can be expressed as

$$2CaCO_3 + SO_2 + 2NO_2 + 6H_2O + O_2 =$$

$$CaSO_4 \cdot 2H_2O + Ca(NO_3)_2 \cdot 4H_2O + CO_2 \tag{6.14}$$

Following Eq. (6.10), the rate of growth of crust with respect to the $SO_2 + NO_2$ reaction can be expressed by

$$\frac{d\delta}{dt} = \frac{M_B}{\rho_B} \frac{b}{a} \left(\frac{\alpha_m \delta}{D_{en} C_{SO_2}^{a_1} C_{NO_2}^{a_2}} + \frac{1}{k_{sn} C_{SO_2}^{a_1} C_{NO_2}^{a_2}} + \frac{1}{h_d C_{SO_2}^{a_1} C_{NO_2}^{a_2}} \right)^{-1} \tag{6.15}$$

Rearranging Eq. (6.15) yields a quadratic equation in δ:

$$\frac{1}{2D_{en}} \delta^2 + \left(\frac{1}{k_{sn}} + \frac{1}{h_d} \right) \delta = 2\alpha_m C_{SO_2}^{a_1} C_{NO_2}^{a_2} \frac{M_B}{\rho_B} t \tag{6.16}$$

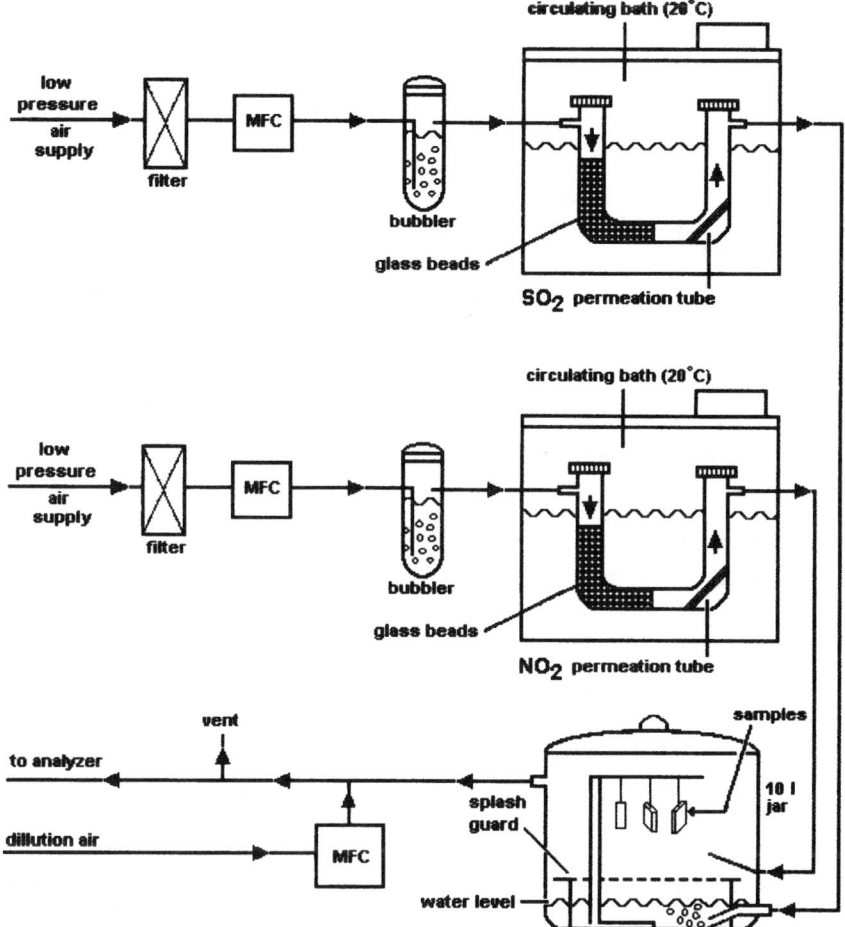

FIGURE 6.5 Chamber for the $SO_2 + NO_2$ reaction. Generation of artificial atmospheres are shown; compare this with the chamber for SO_2 reaction shown in Fig. 5.10. *Source*: Yerrapragada S. S., Chirra, S. R., Jaynes, J. H., Bandyopadhyay, J. K., and Gauri, K. L., Weathering rates of marble in laboratory and outdoor conditions, *J. Environ. Eng.*, **122**(9): 856–863, 1996. Reproduced with the permission of the American Society of Civil Engineers, copyright 1996 ASCE.

where δ is the crust thickness, M_B is the gram molecular weight of calcite (100.9), ρ_B is the density of marble (2.714 g/cm^3), and α_m is the ratio of molar volume of product to reactant (2.1), C_{SO_2} and C_{NO_2} are the concentrations of SO_2 and NO_2 and a_1 and a_2 are their reaction orders. k_{sn} is the rate constant and D_{en} is the virtual diffusivity of Ca^{2+}. For convenience, b/a, the stoichiometric ratio may be taken as 2. Regarding a_1 and a_2, we know that the SO_2 and NO_2 reaction with marble is a first-order reaction and that the measured relative proportion of sulfate and nitrate in the reaction product is 0.7

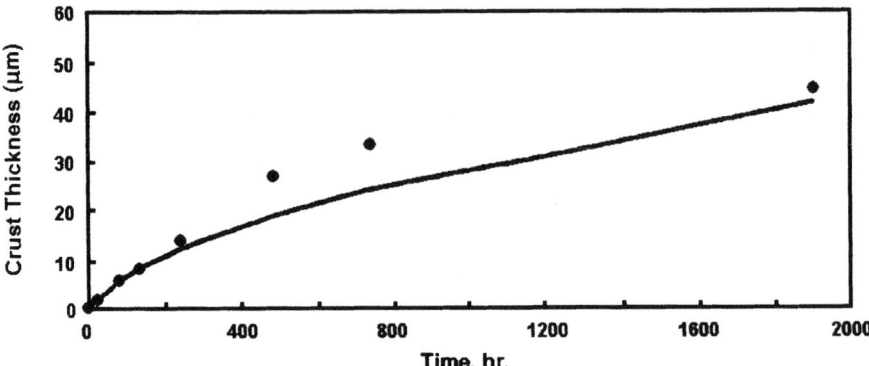

FIGURE 6.6 Marble reaction in 10 ppm each of $SO_2 + NO_2$ atmosphere: application of the shrinking-core model. The curve shows the model-predicted crust thickness after Eq. (6.16) and dots represent actual data points. The regression fit gives values for k_{sn} and D_{en} as 375 cm/h and 0.11 cm²/h. *Source*: Yerrapragada S. S., Chirra, S. R., Jaynes, J. H., Bandyopadhyay, J. K., and Gauri, K. L., Weathering rates of marble in laboratory and outdoor conditions, *J. Environ. Eng.*, **122**(9): 856–863, 1996. Reproduced with the permission of the American Society of Civil Engineers, copyright 1996 ASCE.

and 0.3, respectively. Therefore, the values of a_1 and a_2 must also be 0.7 and 0.3. Figure 6.6 shows the rate curve resulting by the application of Eq. (6.16). The constants k_{sn} and D_{en} determined by fitting the regression curve to the experimental data give the values of 375 cm/h and 0.11 cm²/h, respectively.

6.3.4 Rate of Growth of Crust on Marble in the Louisville Environment

The reason for experiments in concentrated atmospheres, described before, is that adequate data can be collected in a short period of time. Furthermore, the laboratory experiments are in controlled conditions while outdoor conditions can unpredictably change. Nevertheless, we describe below experiments on Carrara and Georgia marble exposed to an ambient environment in Louisville, Kentucky.

Samples were placed at six sites throughout Jefferson County, KY (Fig. 6.7) for up to 20 months. The Jefferson County Air Pollution Control District measured SO_2 and NO_2 concentrations at site 5 and we monitored these gases at the roof of the Natural Sciences Building, University of Louisville. The crust thickness was determined by X-ray photoelectron microscopy and ion chromatography (Chap. 5).

Figure 6.8 gives the regression rate curves for the experimental data for Carrara marble from sites 5 and 6 where the gas concentrations were equal, nearly 10 ppb and 25 ppb SO_2 and NO_2, respectively. The unique values for k_s and D_e for this best-fit curve are 395 cm/h and 0.13 cm²/h. Comparing these values with those determined from laboratory experiments (Table 6.2), one finds excellent agreement between the two.

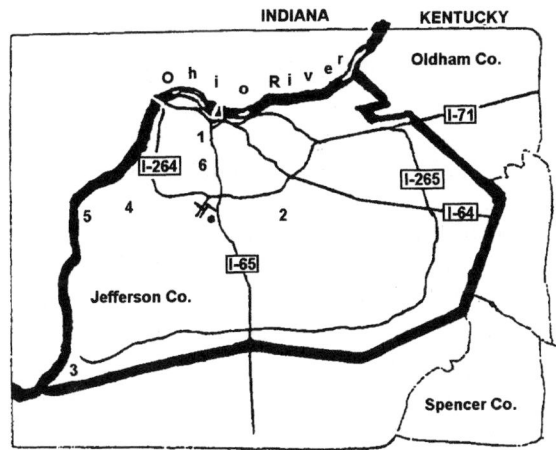

FIGURE 6.7 Map of Jefferson County, Kentucky. Showing sites of exposure: (1) Floyd and Jefferson, (2) Eliahu, (3) convenience store, (4) Wellington, (5) Riverport, and (6) Natural Science Building, University of Louisville. *Source*: Yerrapragada, S., Jaynes, J. H., Chirra, S. R., and Gauri, K. L., Rate of weathering of marble due to dry deposition of ambient sulfur and nitrogen dioxides, *Anal. Chem.*, **66**(5): 655–659, 1994. Reproduced with the permission of the American Chemical Society, copyright 1994 ACS.

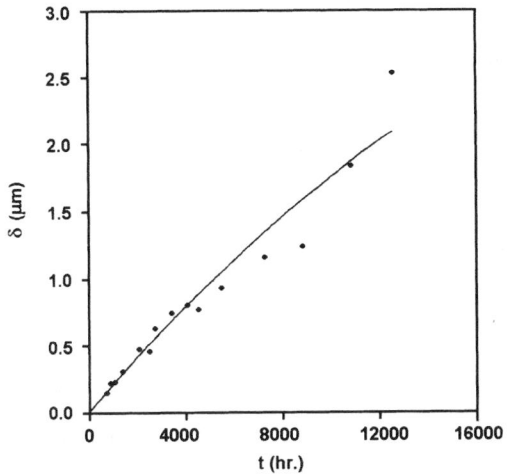

FIGURE 6.8 Carrara marble reaction at sites 5 and 6 (Fig. 6.7). The curve shows the shrinking-core model–predicted crust thickness after Eq. (6.16) and dots represent actual data points. SO_2 and NO_2 concentrations at these sites were nearly 10 ppb and 25 ppb, respectively. The regression fit gives values for k_s and D_e as 395 cm/h and 0.13 cm^2/h. *Source*: Yerrapragada, S., Jaynes, J. H., Chirra, S. R., and Gauri, K. L., Rate of weathering of marble due to dry deposition of ambient sulfur and nitrogen dioxides, *Anal. Chem.*, **66**(5): 655–659, 1994. Reproduced with the permission of the American Chemical Society, copyright 1994 ACS.

The data for other sites in Jefferson County where the concentrations of these gases were not monitored are shown in Fig. 6.9. The curves are drawn using the model values for k_s and D_e, being 395 cm/h and 0.13 cm^2/h. Now, it is possible to calculate the SO$_2$ and NO$_2$ concentrations for these sites since the experimental data on crust thickness are known. These concentrations are found to be 10 ppb to 20 ppb for SO$_2$ and 22 ppb to 32 ppb for NO$_2$. Obviously, these are consistent with the possible ambient levels.

Furthermore, Figure 6.10 shows the crust thickness for samples of Georgia marble exposed next to Carrara marble samples shown in Figure 6.9. It is apparent that the Georgia marble experienced less of a reaction than the Carrara marble. This can be attributed to the different petrofabrics of these two marble varieties, discussed in Chapter 1. The more compact Georgia marble appears to offer greater diffusion resistance while the surface reaction rates of these two marble types should nearly be the same. The calculated D_e for the Georgia marble is found to be 0.068 cm^2/h.

We give a sample calculation for the crust thickness of Carrara marble in the prevailing environment of Louisville using the values of the rate constant and diffusion coefficient as determined by laboratory experiments. These and other rate parameters are given in Table 6.3.

TABLE 6.3   Environmental Conditions and Other Parameters for a Sample Calculation of Crust Thickness on Carrara Marble in Louisville Area

Concentration of SO$_2$ (10 ppb), C_{SO_2}	4.09×10^{-13} mol/cm^3
Concentration of NO$_2$ (25 ppb), C_{NO_2}	10.23×10^{-13} mol/cm^3
Order of reaction, a_1, a_2	0.7, 0.3
Reaction rate constant k_{sn}	375 cm/h
Effective diffusivity D_{en}	0.11 cm^2/h
Mass-transfer coefficient h_d	$\cong \infty$
Molecular weight of marble, M_B	100.09
Density of marble, δ_B	2.714 g/cm^3
Molar volume proportionality constant α_m	2.1
Time of exposure	1 yr

The crust thickness from Eq. (6.16) is

$$\left(\frac{1}{2 \times 0.11 \times 4.09 \times 10^{-13}} \right) \delta^2 + \left(\frac{1}{375 \times (4.09 \times 10^{-13})^{0.7} (10.23 \times 10^{-13})^{0.3}} \right) \delta$$

$$= 2 \times 2.1 \frac{100.09 \times 365 \times 24}{2.714}$$

Taking the positive root we get

$$\delta = 1.92 \ \mu m$$

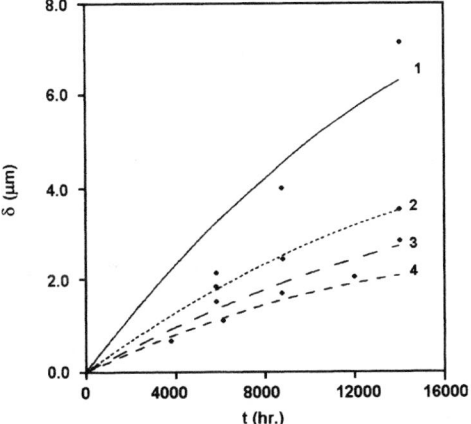

FIGURE 6.9 Carrara marble reaction at sites 2 to 4 (Fig. 6.7) (gas concentration not monitored). The curve shows the shrinking-core model–predicted crust thickness after Eq. (6.16) and dots represent actual data points. The application of rate constants in Fig. 6.8 gave the SO_2 and NO_2 concentrations as 10–20 ppb and 22–32 ppb, respectively. The plot of site 1 shows somewhat aberrant behavior. This site is close to a freeway and a power plant. *Source*: Yerrapragada, S., Jaynes, J. H., Chirra, S. R., and Gauri, K. L., Rate of weathering of marble due to dry deposition of ambient sulfur and nitrogen dioxides, *Anal. Chem.*, **66**(5): 655–659, 1994. Reproduced with the permission of the American Chemical Society, copyright 1994 ACS.

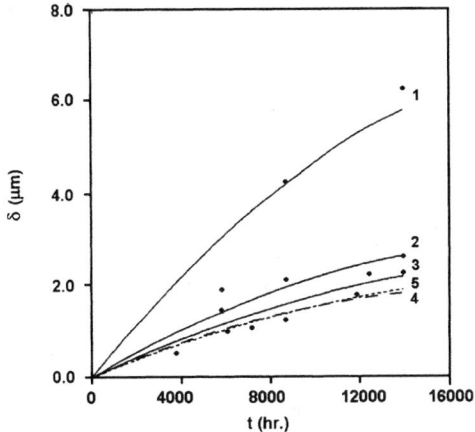

FIGURE 6.10 Georgia marble reaction at sites 1 to 5 (Fig. 6.7). The curve shows the shrinking-core model–predicted crust thickness after Eq. (6.16) and dots represent actual data points. SO_2 and NO_2 concentrations at site 5 were nearly 10 ppb and 25 ppb, respectively. This marble shows lesser reactivity than Carrara marble. The average calculated D_e for this marble is $0.0.068 \text{ cm}^2/\text{h}$ as compared with $0.13 \text{ cm}^2/\text{h}$ for Carrara marble. *Source*: Yerrapragada, S., Jaynes, J. H., Chirra, S. R., and Gauri, K. L., Rate of weathering of marble due to dry deposition of ambient sulfur and nitrogen dioxides, *Anal. Chem.*, **66**(5): 655–659, 1994. Reproduced with the permission of the American Chemical Society, copyright 1994 ACS.

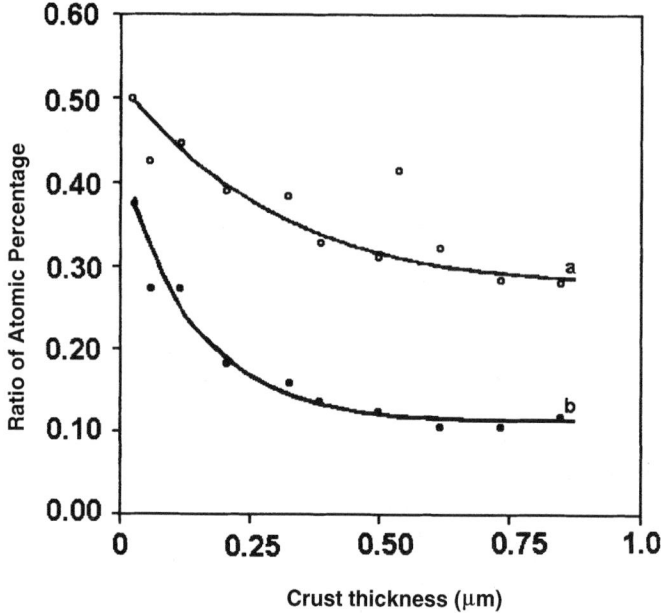

FIGURE 6.11 Crust thickness on Georgia marble measured by sputter profiling (Fig. 5.13). Samples were exposed for 6 months at site 1 (Fig. 6.7). The crust thickness is nearly 0.6 μm thick. *Source*: Yerrapragada, S., Jaynes, J. H., Chirra, S. R., and Gauri, K. L., Rate of weathering of marble due to dry deposition of ambient sulfur and nitrogen dioxides, *Anal. Chem.*, **66**(5): 655–659, 1994. Reproduced with the permission of the American Chemical Society, copyright 1994 ACS.

This thickness, obtained by using the model equation, is somewhat larger than that measured for Georgia marble (Fig. 6.11) for which it is nearly 0.6 μm for a six-month exposure. However, as we have noted before, Georgia marble is less reactive than Carrara marble. The programs at the ftp site (Appendix C) can be used to calculate the crust thickness in any atmospheric composition of SO_2 and NO_2.

This discussion leads to the conclusion that the rate constants based on accelerated laboratory experiments are valid for evaluation of outdoor weathering. Thus, from the application of the model equations one can predict either the rate of growth of crust or the ambient SO_2 and NO_2 concentrations. However, we have noticed that the magnitude of reaction can vary considerably from sample to sample even when exposed simultaneously both in indoor as well as outdoor conditions and that the variations may be as large as an order of magnitude.

6.3.5 Model for Recession of Rain-Washed Surfaces

The mechanism governing the recession of marble surfaces exposed to rain are highly complex and cannot be simulated in the laboratory. Further, the rate of recession is much faster than the rate of growth of crust, so that it is feasible to model the surface recession on the basis of data collected from outdoor experiments. Later we will describe experiments that we performed outdoors to determine the rate of surface recession, but first we will discuss the phenomenon of acid rain that causes carbonate stone in polluted environments to erode at a phenomenal rate. Figure 6.12 shows the distribution of acid rain in the United States.

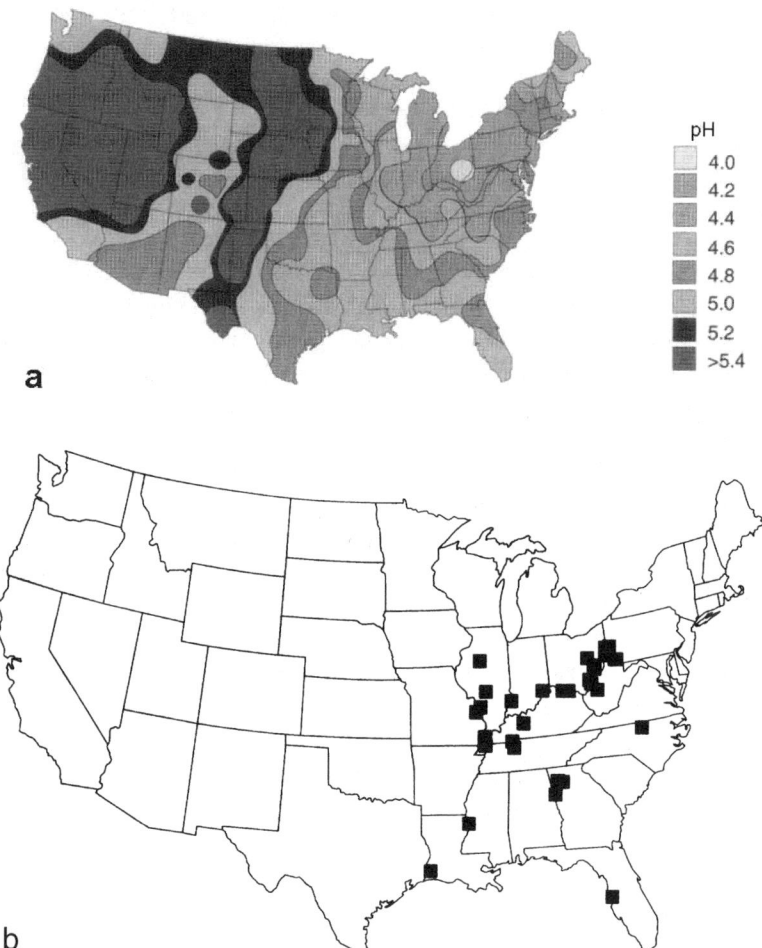

FIGURE 6.12 Distribution of acid rain in the United States. Notice that the entire country experiences acid rain (a), defined as rain, or any form of precipitation, with a pH lower than 5.6. Acid rain forms by the solution in rainwater of sulfur dioxide (SO_2) and nitrogen oxide (NO + NO_2) gases released into the atmosphere in large quantities in industrial countries. In the United States the concentration of major power plants along the Ohio Valley (b) produces the most acidic rain. *Source*: Adapted from a 1998 map from the National Atmospheric Deposition Program (NSRP-3).

Rain, in polluted as well as nonpolluted environments, becomes saturated with CO_2 as it passes through the atmosphere. When it is in equilibrium with the current (nearly 360 ppm) partial pressure of CO_2 in the atmosphere, rain has a pH of nearly 5.6 and is often termed "clean rain," even though it is somewhat acidic. Such rains fall, at present, in nonpolluted environments only. In polluted environments, however, SO_2 and NO_2 also dissolve in rainwater, lowering the pH to below 5.6, which is known as acid rain and is defined as rain with a pH below 5.6. The major constituents of acid rain are carbonic acid (H_2CO_3), sulfuric acid (H_2SO_4), and nitric acid (HNO_3). These acids attack calcite ($CaCO_3$) through H^+ ions (Eqs. 4.1, 4.2, 5.9) to produce soluble Ca^{2+}, HCO_3^-, and CO_3^{2-} ions. The sulfuric acid and nitric acid, in addition, produce water-soluble salts such as gypsum ($CaSO_4 \cdot 2H_2O$) and nitrocalcite [$Ca(NO_3)_2 \cdot 4H_2O$] (Eqs. 5.6 through 5.11), which drain away as Ca^{2+}, SO_4^{2-}, and NO_3^{2-} ions. Thus the surface recession can be measured as loss of calcium ions converted to parent calcium carbonate by mass balance.

Surface recession calculated from loss of Ca^{2+} ions gives the overall acid rain effect. However, to model the surface recession so that it can be predicted from any environmental condition, it is necessary to consider the effect of each component contributing to the overall recession. Further, we must also know the contribution of dry deposition during periods between rainfall events. In the following paragraph we describe experiments done at the roof of the Natural Science Building, University of Louisville (Site 6, Fig. 6.7), where we collected pertinent data for modeling surface recession.

Two sets of marble samples were exposed: one beneath a canopy to yield the dry deposition effect, and the other nearby, in the open, tied to stands, and placed above funnels so that the rain runoff over the samples could be collected. A third set of samples, consisting of porcelain plates, was also exposed outside the shelter, beside the funnels, to account for the nonreactive sulfate and nitrate particles deposited during the dry deposition periods. After exposure, the sheltered samples were leached in deionized water. This water, as well as the rain runoff over the samples, was analyzed for Ca^{2+}, SO_4^{2-}, and NO_3^- ions; calcium by atomic absorption spectrophotometry; and sulfate and nitrate by ion chromatography. From these data we can derive information regarding the individual contribution of each mechanism to the ionic mass leached as follows.

1. We found that the runoff contains more sulfate (SO_4^{2-}) and nitrate (NO_3^-) than leachate from sheltered samples; the excess sulfate and nitrate must be a result of the corrosive effect of sulfuric acid (H_2SO_4) and nitric acid (HNO_3) in the rain. In the model equation for surface recession, Eq. (6.17), the quantities of sulfate and nitrate in the runoff, converted to an equivalent mass of gypsum and nitrocalcite, are expressed by the rate constant k_r obtained in the same manner as k_s (Eq. 6.10) in the case of dry deposition by the regression of the runoff data.

2. We found further that the runoff contains a much larger mass of Ca^{2+} than needed to balance the mass of sulfate and nitrate. The excess Ca^{2+}, represented by δ_e in Eq. (6.17), can be attributed to the following:

- *H^+ Effect of "Clean Rain."* The clean rain as indicated above has a pH of 5.6, which is due to the presence of carbonic acid (H_2CO_3), which enhances the dissolution of calcite over its solubility in pure water (0.014 g/l). The effect of clean rain is represented by the term N_d in Eq. (6.18).

- *H^+ Ion Effect of Acid Rain.* The sulfuric (H_2SO_4) and nitric (HNO_3) acids in the rainwater produce a low pH (< 5.6), resulting in a further increase of the dissolution of calcite. This is shown as a function of H^+ activity in Eq. (6.18).

TABLE 6.4   Surface Recession and Other Parameters for Outdoor Experiments

Sample Number (1)	Surface Area (cm^2) (2)	Dry-Deposition Period (h) (3)	Rainfall (mm) (4)	A^a (5)	B^b (6)	C^c (7)	D^d (8)
				Surface Recession (μm)			
1	57	72	27	0.04	0.00	0.12	0.16
2	253	96	13	0.01	0.02	0.10	0.13
3	265	120	10	0.01	0.02	0.08	0.11
4	250	120	23	0.01	0.03	0.18	0.22
5	48	144	38	0.14	0.01	0.32	0.47
6	256	144	10	0.03	0.02	0.06	0.11
7	53	192	19	0.01	0.02	0.20	0.23
8	47	192	25	0.07	0.02	0.20	0.29
9	57	264	82	0.10	0.00	0.65	0.75
10	252	288	10	0.03	0.04	0.08	0.15
11	56	312	17	0.12	0.05	0.14	0.31
12	51	336	36	0.11	0.04	0.28	0.43
13	55	408	19	0.04	0.01	0.23	0.28

Source: Yerrapragada, S. S., Chirra, S. R., Jaynes, J. H., Li, S., Bandyopadhyay, J. K., and Gauri, K. L., Weathering rates of marble in laboratory and outdoor conditions, *J. Environ. Eng.*, **122**(9): 856–863, 1996.

Note: Surface recession was calculated as follows:
[a]Calculated form sulfates and nitrates in runoff.
[b]Calculated from sulfates and nitrates extracted from samples after rain event (these are the sulfates and nitrates that percolated into the sample and were retrieved by repeated leaching of the sample in deionized water).
[c]Calculated from calcium in runoff in excess of that needed to balance sulfates and nitrates.
[d]Calculated from total calcium in runoff. Note that D^d is the sum of A^a, B^b, and C^c.

The surface recessions due to these various mechanisms are given in Table 6.4. Surface recession is also affected by the volume of rain washing the sample surface, leading to the following final expression:

$$\delta_m \left(\frac{\mu m}{m} \right) = \delta_e + 2 \frac{k_r M_B C_{SO_2}^{0.7} C_{NO_2}^{0.3} t}{V \rho_B} \quad (6.17)$$

where δ_m ($\mu m/m$) is the surface recession of marble per meter of rain; k_r (cm/h) is the rate constant expressing the effects of dry deposition and sulfuric and nitric acids in rain; V is the rainfall amount (m), and the expression for δ_e ($\mu m/m$) is given by

$$\delta_e = N_d + f(H^+) \quad (6.18)$$

Now, we will determine the values of k_r, N_d, and $f(H^+)$. Regression analysis of the data on the total surface recession expressed by Ca^{2+} in runoff (Table 6.3, column 8) for times (h) of rainfall in individual events (column 4) gives the rate curve plotted in Figure 6.13. The value of the rate constant k_r was found to be 2452 cm/h. This value represents the dry-deposition effect in addition to the attack of H_2SO_4 and HNO_3 in the rain as expressed by the sulfate and nitrate in the runoff. The value of δ_e is shown by the intercept of the regression curve as being 7.6 $\mu m/m$ and results from the combined effect of CO_2 and excess H^+ ions in the low pH (4.5) of the Louisville rain. We were able to calculate the individual contributions to δ_e using a plot of marble solubility as a function of H^+-ion loading predicted by the equilibrium model

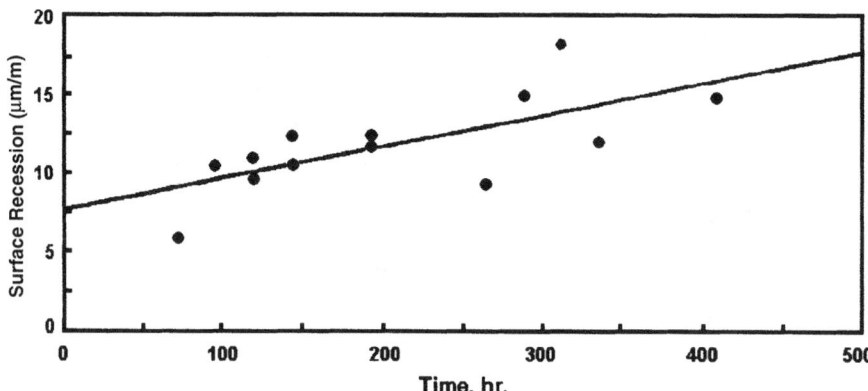

FIGURE 6.13 Rate of surface recession. The curve was developed by the application of Eq. (6.21) and dots represent surface recession calculated on the basis of calcium in the rain runoff (Table 6.1, column D) over sample surfaces. The samples were from Carrara marble exposed on the top of Natural Science Building (Fig. 6.7, site 6). *Source:* Yerrapragada S. S., Chirra, S. R., Jaynes, J. H., Bandyopadhyay, J. K., and Gauri, K. L., Weathering rates of marble in laboratory and outdoor conditions, *J. Environ. Eng.*, 122(9): 856–863, 1996. Reproduced with the permission of the American Society of Civil Engineers, copyright 1996 ASCE.

PHREEQEE (Reddy and Baedecker). According to that plot, an average rate of Ca^{2+} production at 25°C can be given by

$$Ca^{2+} \text{ (mmol/l)} = 0.18 + 0.75 \text{ H}^+ \tag{6.19}$$

where the intercept 0.18 represents effect of CO_2 and the H^+ term indicates hydrogen-ion loading at pH in the range of 3 to 5.5. Expressed as marble recession (δ_e) by converting Ca^{2+} (mmol/l) to $CaCO_3$ (μm), Eq. (6.19) becomes

$$\delta_e \text{ (μm/m)} = 6.56 + 27.38 \text{ H}^+ \tag{6.20}$$

Finally, the total surface recession per meter of rain may be given by

$$\delta_m \text{ (μm/m)} = 6.56 + 27.38 \text{ H}^+ + \frac{1.131 \times 10^{13} \, C_{SO_2}^{0.7} C_{NO_2}^{0.3} t}{V} \tag{6.21}$$

where H^+ is in mmol/l, C_{SO_2} and C_{NO_2} are in mol/cm³, t is in yr, and V is in m.
  Using Eq. (6.21) and a pH of 4.5, we get

$$\delta_m \text{ (μm/m)} = 6.56 + 27.38(10^{-4.5} \times 1000)$$

$$+ \frac{1.131 \times 10^{13} \times (4.09 \times 10^{-13})^{0.7} \times (10.23 \times 10^{-13})^{0.3} (365 \times 24)}{1.13}$$

$$= 12.82 \text{ μm/m}$$

Thus the rate of surface recession in Louisville for the 1.13 m annual rainfall is calculated to be 14.48 μm/y.

Nevertheless, more study is needed to confirm the large contribution of carbon dioxide to surface reduction in a polluted environment shown by the Louisville example. This concern is based on the perception that the CO_2 level of the atmospheric composition has not changed much lately in industrial countries, yet the surface reduction has greatly increased. Apparently the effect of CO_2 is overstated by our model equations.

We will now compare some features of surface recession and crust formation and give their relative contribution to stone decay and their influence on conservation measures. In early phases of rain exposure, leaching causes some faintness of relief but in prolonged exposure calcite grains tend to fall off due to solution around grain boundaries. At this later stage the inscriptions become illegible and the relief loses detail yet preserves the original form. This contrasts with the exfoliation of the crust where at such an advanced state of decay the sculptured form becomes totally obliterated (Fig. 5.3).

Further, the solution of calcite along grain boundaries can be helpful for conservation treatment by allowing preservatives to penetrate inside the stone. This contrasts with weathering at sheltered sites where impermeable crusts impede penetration of consolidants in attempts to bond the crust with the substrate.

6.4 DEPOSITION-VELOCITY MODEL

The deposition velocity v_d (cm/s) is the overall mass-transfer coefficient, which is equal to the average rate of migration of a dilute gas towards an absorbing surface. Thus, it is the reverse of the overall aerodynamic resistance, R_1 (s/cm). Flux F (mol·cm^{-2}·s^{-1}), the rate of transfer, can be conveniently expressed as

$$F = \frac{C_{Ab}}{R_1} \qquad (6.22)$$

where C_{Ab} is the concentration of SO_2 (mol/cm^3) in the air. Thus,

$$F = C_{Ab}v_d \qquad (6.23)$$

The deposition velocity can be determined by measuring the difference in the concentration of reactant gas at the inlet of reactor and that found at the outlet, or it can be estimated from the rate of uptake of gas by the reactive surface. In the first case,

$$F = \frac{m_{SO_2,in} - m_{SO_2,out}}{M_{SO_2}A_0} = v_d C_{Ab} \qquad (6.24)$$

where $m_{SO_2,in}$ and $m_{SO_2,out}$ are the inlet and outlet mass flow rate of SO_2, M_{SO_2} is the sulfur dioxide molecular weight, and A_0 is the external surface area of the sample.

The surface flux can be related to the rate of the reactant mass consumed or the reaction product formed per surface area through conservation of mass and stoichiometry ratio:

$$\frac{1}{A_0}\frac{dm_P}{dt} = \alpha M_P \qquad (6.25)$$

$$\frac{d\delta}{dt} = \alpha \frac{M_P}{\rho_P} v_d C_{Ab} \qquad (6.26)$$

where m_P, M_P, and ρ_P are the mass, molecular weight, and density of the product. Equation (6.26) can be integrated to give the crust thickness:

$$\delta = \alpha \frac{M_P}{\rho_P} v_d C_{Ab} t \qquad (6.27)$$

Figure 6.14 shows results of an experiment in which marble slabs were exposed in a 10 ppm SO_2 atmosphere with a relative humidity less than or equal to 100%. It is apparent that v_d varied between 0.06 cm·s^{-1} and 0.01 cm·s^{-1} and in longer exposure, there was a decrease in the deposition velocity, as expected.

The deposition velocity is a widely used parameter for describing the uptake of gases by surfaces. However, it has the drawback that it is a bulk coefficient, which accounts for both the surface effect and the aerodynamic mass-transfer effect.

Surface effects, such as the possible wetness of the surface, can increase deposition velocity over an order of magnitude. Covering of marble surfaces with gypsum after the initial surface reaction reduces the deposition velocity because the buffering effect of carbonate minerals is, in fact, responsible for high deposition velocities in carbonate rocks.

Furthermore, these surface effects, combined with the roughness of surface, change the aerodynamic mass transfer. Deposition velocity, though a useful parameter to describe the change of a solid, is valid only under instantaneous specific conditions of the gas–solid environment. For considerations of crust formations over extended periods of time, it has a limited use.

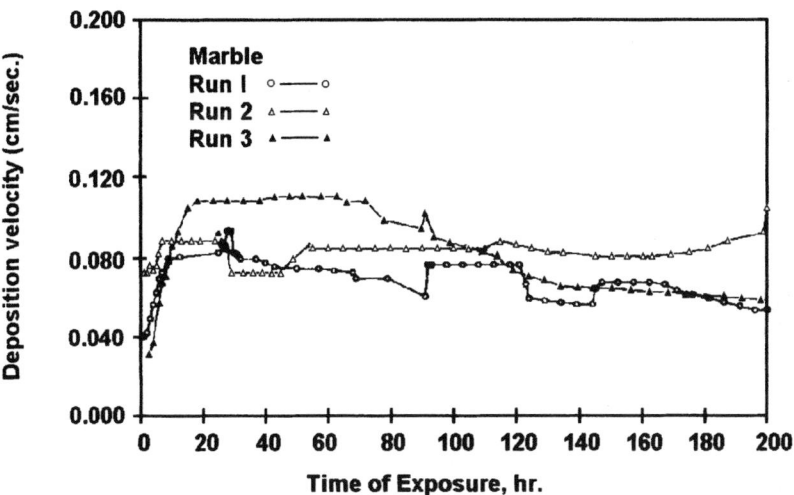

FIGURE 6.14 Sulfur dioxide deposition velocity on marble. The measured deposition velocity was relatively constant except the initial period of 5 h to 10 h, slightly diminishing throughout the exposure period. *Source*: Cobourn W. G., Gauri, K. L., Tambe, S., Li, S., and Saltik, E., Laboratory measurements of sulfur dioxide deposition velocity on marble and dolomite surfaces, *Atm. Environ.*, **27**B(2): 193–201, 1993. Reproduced with permission of Elsevier Science, copyright 1993 Elsevier.

6.5 PORE MODELS

Dolomite and limestone are generally quite porous. Their reaction with SO_2, etc., perceptibly changes their porosity. Therefore, in addition to the shrinking-core model, which accounts for the change in the reactants or the reaction products, the conversion can be described through the change in porosity they undergo due to reaction. The models used for this are commonly termed *percolation models*. We will discuss below two such models, namely, the distributed-pore model (Zarkinitis and Sotirchos, 1989) for limestone and the random-pore model (Bhatia and Perlmutter, 1980) for dolomite; needless to say these models can be used interchangeably. We determined the change in porosity by dissolving the reaction product in deionized water from samples that had been exposed in laboratory experiments.

6.5.1 Distributed-Pore Model

The pore network in a porous medium can be described by the pore-size distributions and the coordination number (Z) defined as the number of pores emanating from a site or number of pores accessible in the network. The accessible porosity of a Z coordinated network can be written as

$$\phi^A = \phi\left[1 - \left(\frac{\phi^R}{\phi}\right)^{(2Z-2)/(Z-2)}\right]$$

(6.28)

where ϕ and ϕ^A are total and accessible porosity, and ϕ^R can be obtained from the following relationship:

$$\phi^R(1 - \phi)^{Z-2} - \phi(1 - \phi)^{Z-2} = 0$$

(6.29)

Equation (6.29) is valid when $\phi > \phi_c$, where

$$\phi_c = \frac{1}{Z - 1}$$

(6.30)

Furthermore, the relationship of accessible surface area a^A to the pore structure can be evaluated by

$$a^A = K[\phi(1 - \phi) - \phi'(1 - \phi')]$$

(6.31)

where K, a form of coefficient, is obtained by matching the initial surface area to the accessible surface area and ϕ' is the isolated porosity, obtained from the difference between the total porosity ϕ and the accessible porosity ϕ^A and

$$K = e^{1/(\phi-\phi_c)} \tag{6.32}$$

Thus the pore parameters and the reaction rate can be related as

$$\frac{d\phi}{dt} = \frac{M_B}{\rho_B} k_s C_{SO_2} a^A \tag{6.33}$$

where M_B and ρ_B are the molecular weight and density of the reactant and k_s and C_{SO_2} are obtained as shown previously. Finally, conversion expressed as mass consumed per unit mass of solid reactant can be calculated from the following equation:

$$X = \frac{\phi - \phi_0}{1 - \phi_0} \tag{6.34}$$

We will now apply the distributed-pore model to various varieties of limestone discussed previously. As seen from Eqs. (6.33) to (6.34), the elements needed to apply this model are the surface rate constant and pore-structure properties, such as porosity, coordination number, and the related K factor.

The initial porosity ϕ_0, the end porosity, and the bulk density of the limestone varieties are shown in Table 6.5. While the rate constant, k_s, was determined from the

TABLE 6.5 Porosity and Density of the Limestone Varieties[a]

Sample Name	Initial Porosity (%)	Final Porosity (%)[b]	Density (g/cm³)
Cordova Cream	26.13	27.50	1.97
Cordova Shell	21.65	22.67	2.08
Lueders	19.94	21.23	2.14
Bedford	13.74	14.56	2.29

Source: Bandyopadhyay, J. K., Annamali, S., and Gauri, K. L., Application of artificial neural networks in modeling limestone–SO$_2$ reaction, *AIChE J.*, **42**(8): 2295–2306, 1996.

[a]For a description of these limestone varieties see Chapter 1.

[b]The final porosity was measured after the reaction product had been leached out.

regression analysis of the experimental rate data, K was calculated using the total and critical porosity [Eq. (6.30)]. The calculated conversion-time plots based on Eq. (6.34) are shown in Figure 6.15, in which a good match between these and observed conversions is seen, suggesting the usefulness of percolation models in evaluating SO_2–limestone reactions. It should be noted, though, that the Z value that gave the best fit to the observed conversion of limestone in all varieties reported ranged from 10 to 14. This value is higher than that conventionally used in percolation models, but seems to be valid considering the complex network of pores in limestone.

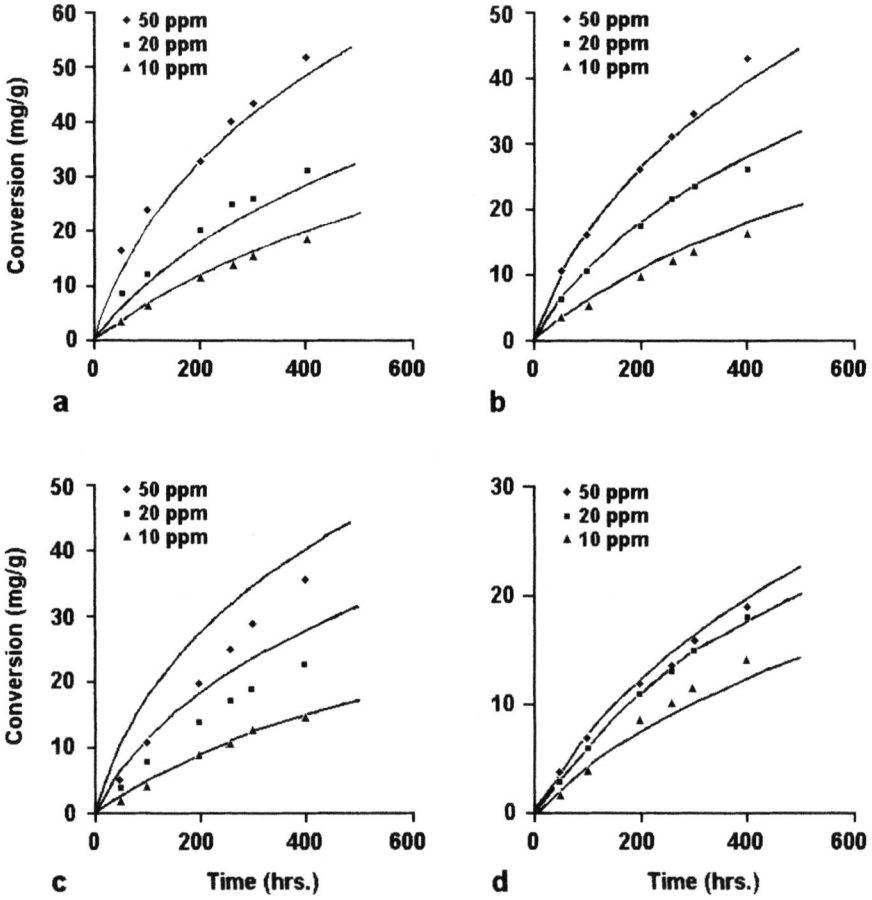

FIGURE 6.15 Limestone reaction in variable SO_2 concentrations: application of the distributed-pore model. Curves show the predicted crust thickness after Eq. (6.35) and dots represent actual data points. Limestone types are (a) Cordova Cream; (b) Cordova Shell; (c) Lueders; and (d) Bedford. *Source*: Bandyopadhyay, J. K., Annamalai, S, and Gauri, K. L., Application of artificial neural networks in modeling limestone–SO_2 reaction, *Am. Inst. Chem. Eng. J.*, **42**(8): 2295–2302, 1996. Reproduced with the permission of the American Institute of Chemical Engineers, copyright 1996 AIChE.

6.5.2 Random-Pore Model

The random-pore model, though based upon pore surface area, considers the change in pore structure and volume over time. For instance, it is assumed that the pores are initially nonoverlapped. With increasing time of reaction the pores enlarge, yielding a larger surface area, which then enhances the reaction. In a continued reaction, the pores may begin to overlap so that the surface area is reduced and the reaction decelerates.

According to Bhatia and Perlmutter (1980) the surface area and pore volume in an nonoverlapped pore phase can be described as

$$S_N = \sqrt{S_{N0}^2 + 4\pi L_{N0}(V_N - V_{N0})} \tag{6.35}$$

$$V_N = V_{N0} + S_{N0}(k_r C_{AS}^n t) + \pi L_{N0}(k_r C_{AS}^n t)^2 \tag{6.36}$$

which for a general overlapped and nonoverlapped system appears as

$$V_O = 1 - (1 - V_{N0}) \exp\left(\frac{k_r C_{AS}^n}{(1 - V_{N0})}(S_{N0} + \pi L_{N0} k_r C_{AS}^n t)\right) \tag{6.37}$$

where S_N (cm^2/cm^3) and V_N (cm^3/cm^3) represent the pore surface area and the pore volume at any instant of time, and S_{N0} (cm^2/cm^3), L_{N0} (cm/cm^3), and V_{N0} (cm^3/cm^3), respectively, represent the initial pore surface area, pore length, and pore volume of the nonoverlapped cylindrical pore system, normalized by the volume of the particle. The quantity V_O (cm^3/cm^3) is the pore volume of the overlapped pore system at any instant; k_r, n, C_{AS}, and t denote the surface reaction rate, the order of the reaction, the surface concentration of the gaseous reactant (mmol/cm^3), and time (h), respectively. Thus, the generalized rate expression is given by

$$S_O = S_{N0} \frac{1 - X}{(1 - \tau/\sigma)^3} \sqrt{1 - \Psi \ln\left[\frac{(1 - X)}{(1 - \tau/\sigma)^3}\right]} \tag{6.38}$$

where S_O (cm^2/cm^3) is the pore surface area of the overlapped pore system at any time t; and X represents the solid conversion, defined as mass of reactant consumed per unit mass of solid reactant. The structural parameter (Ψ), the particle size parameter (σ), and the dimensionless time (τ) are defined as

$$\Psi = \frac{4\pi L_0 (1 - \varepsilon_0)}{S^2 N_O} \tag{6.39a}$$

$$\sigma = \frac{R_0 S_{N0}}{1 - \varepsilon_0} \tag{6.39b}$$

$$\tau = \frac{k_r C_{AS}^n S_{N0} t}{1 - \varepsilon_0} \tag{6.39c}$$

where, R_0 is the initial radius (cm), and ε_0 denotes the porosity of unreacted solid, defined as the pore volume divided by the bulk volume of the solid, equal to V_{NO}. The pore parameters L_{N0}, S_{N0}, and ε_0 are calculated as follows:

$$L_{N0} = \frac{1}{\pi} \int_0^\infty \frac{\nu_0(r)}{r^2} \, dr \tag{6.40}$$

$$S_{N0} = 2 \int_0^\infty \frac{\nu_0(r)}{r} \, dr \tag{6.41}$$

$$\varepsilon_0 = \int_0^\infty \nu_0(r) \, dr \tag{6.42}$$

where $\nu_0(r)$ represents the pore-size distributions that can be obtained from mercury porosimetry.

Combining Eqs. (6.35) through (6.42), the conversion–time relationship for a spherical particle is obtained as

$$X = 1 - \left(1 - \frac{\tau}{\sigma}\right)^3 \exp\left[-\tau\left(1 + \frac{\varphi\tau}{4}\right)\right] \tag{6.43}$$

We will now apply the random pore model to conversion of Laurel dolomite (Chaps. 1 and 5). The general SO_2–dolomite reaction is given by

$$CaMg(CO_3)_2 + 2SO_2 \xrightarrow{H_2O, O_2} CaSO_4 \cdot 2H_2O + MgSO_4 \cdot 7H_2O + 2CO_2 \tag{6.44}$$

We have explained the formation of gypsum ($CaSO_4 \cdot 2H_2O$) and epsomite ($MgSO_4 \cdot 7H_2O$) as the reaction products and their rate of formation in Chapter 5. For the application of the random-pore model, we dissolved these salts away by rinsing the samples with deionized water and at different stages of the reaction determined

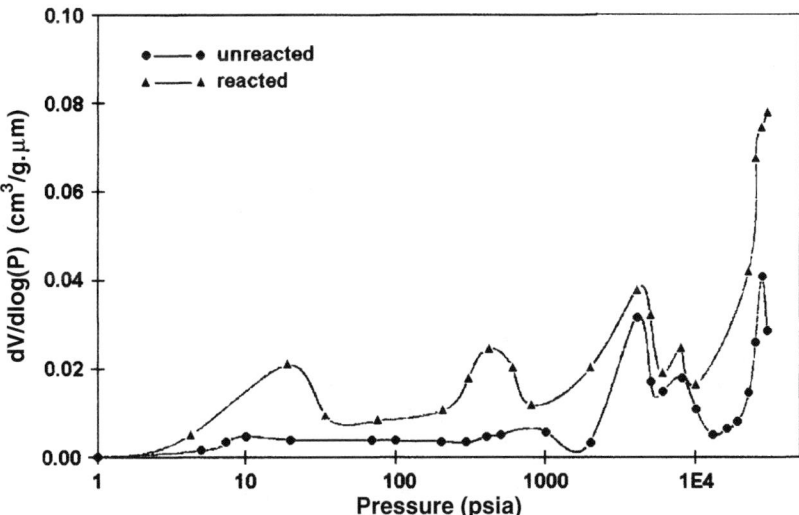

FIGURE 6.16 Porosity change in dolomite, showing pore-size distributions in fresh and reacted samples of Laurel dolomite. In the case of reacted samples, crust had been removed by leaching before determination of pore-size distributions. *Source*: Tambe, S. S., Yerrapragada, S. S., and Gauri, K. L., Kinetics of SO₂-Dolomite Reaction: Application of random Pore Model, *J. Mater. Civil Eng.*, **6**(1): 65–77, 1994. Reproduced with the permission of the American Society of Civil Engineers, copyright 1994 ASCE.

porosity by mercury porosimetry (Chap. 8). Figure 6.16 shows the pore-size distributions before and after the reaction.

Applying Eq. (6.43) to the rate data fitted with the nonlinear optimization routine (Marquardt, 1963) resulted in the values for n and k_r as 0.66 and 10^{-6} $cm^{2.98}/mmol^{0.66}\cdot h$, respectively. In Table 6.6 we summarize the kinetic and structural parameters of the Laurel dolomite related to the random-pore model. The use of these parameters give the predicted curves shown in Fig. 6.17.

6.6 ARTIFICIAL NEURAL NETWORKS MODEL

The theory of artificial neural networks (ANN) is presented in Appendix A. Here, we will describe the application of ANN to predict SO_2–limestone reactivity. The chosen network architecture (Fig. 6.18) has an input layer with three neurons, one hidden layer with five neurons, an output layer with one neuron, and a target value. Porosity, density, and the exposure condition (ppm · h) are the elements of the input layer designated as $I_{1,p}$, $I_{2,p}$, and $I_{3,p}$ in Figure 6.18. As a first step in the operation of ANN these values are multiplied by randomly selected weights ($W_{i,j}$, $i = 1, \ldots, 5; j = 1, \ldots, 3$), summed and processed in the hidden layer to yield data for the subsequent layer, which is the output layer. The value in the output layer is then compared with the

TABLE 6.6 Kinetic and Structural Parameters to Predict Porosity Change in Laurel Dolomite

Surface concentration of SO_2, C_{As}	4.56×10^{-7} mmol/cm^3
Order of reaction, n	0.66
Surface rate constant, k_r	4.9×10^{-6} cm$^{2.98}$/mmol$^{0.66}$
Initial pore length, L_{N0}	4.01×10^{10} cm/cm^3
Initial pore surface area, S_{N0}	1.3×10^5 cm^2/cm^3
Initial particle radius, R_0	0.41 cm
Structural parameter Ψ	26.71
Particle size parameter σ	5.89×10^4
Dimensionless time τ	0.018

Source: Tambe S. S., Yerrapragade, S. S., and Gauri, K. L., Kinetics of SO_2-dolomite reaction: Application of random pore model, *J. Mater. Civil Eng.*, **6**(1): 65–77, 1994. Reproduced with the permission of the American Society of Civil Engineers, copyright 1994 ASCE.

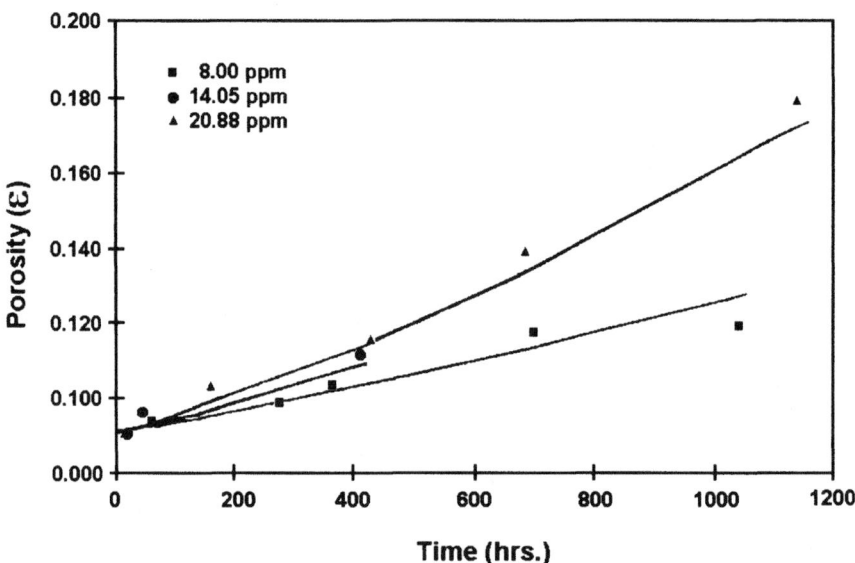

FIGURE 6.17 Porosity change by reaction of dolomite in variable SO_2 concentrations: application of the random-pore model. The model-predicted curve by Eq. (6.45) and experimental points are shown. *Source*: Tambe, S. S., Yerrapragada, S. S., and Gauri, K. L., Kinetics of SO_2–Dolomite reaction: Application of random pore model, *J. Mater. Civil Eng.*, **6**(1): 65–77, 1994. Reproduced with the permission of the American Society of Civil Engineers, copyright 1994 ASCE.

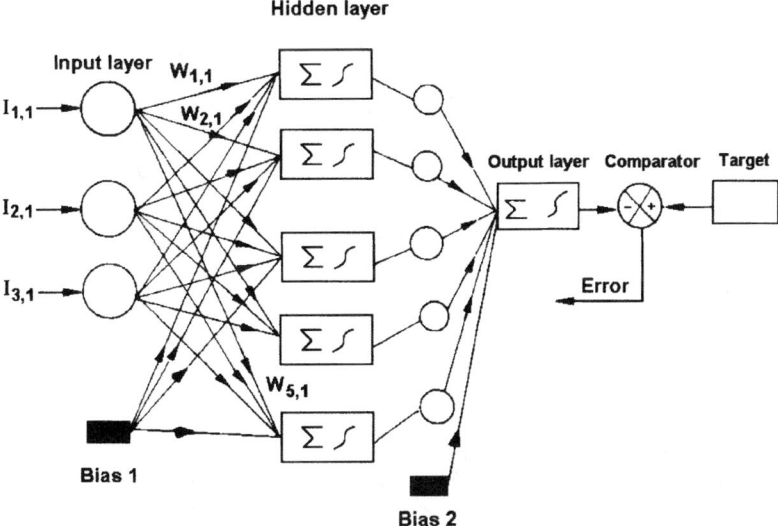

FIGURE 6.18 Architecture of artificial neural network applied to limestone reactions. *Source*: Bandyopadhyay, J. K., Annamalai, S, and Gauri, K. L., Application of artificial neural networks in modeling limestone–SO2 reaction, *Am. Inst. Chem. Eng. J.*, **42**(8): 2295–2302, 1996. Reproduced with the permission of the American Institute of Chemical Engineers, copyright 1996 AIChE.

target, which is the observed conversion. This input–target pair forms what is called a pattern in ANN terminology.

Now let us train ANN to three [(Cordova Cream (CR), Laurel dolomite (LD), and Bedford limestone (BD)] of the four limestone varieties discussed previously. As a first step in the training, the chosen weights are multiplied and the network is further trained by updating weights to minimize the error, that is, the difference between the output layer and the target. This process is repeated until the error reaches a significantly low predefined value. During training, the learning coefficient η and the momentum correction factor α are kept constant at 0.3 and 0.05, respectively.

Figure 6.19 shows the data as well as the rate curves predicted by ANN for four limestone types including Bedford limestone, which was not used in the training process. The predicted rate curves show good matches for all limestone types.

We can further check the validity of the trained network by predicting conversion in ambient SO_2 concentration. For example, the porosity and density of the Bedford limestone are known as well as the SO_2 concentration in the Louisville area. ANN gives a nearly 2-μm-thick crust for one year of exposure in the prevailing 25 ppb SO_2. Now let us compare this with the crust on marble that we measured by XPS (Fig. 6.10), which was found to grow nearly 0.6 μm thick in six months. We know that the Bedford limestone has a somewhat faster reaction rate than the marble. So, the calculated value of a 2-μm-thick crust appears to be a reasonable prediction.

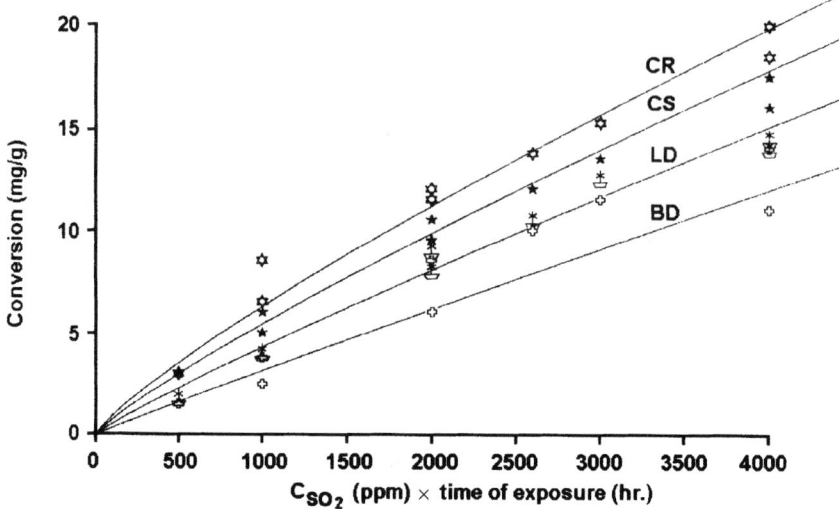

FIGURE 6.19 Limestone reaction in variable SO_2 concentrations: application of artificial neural networks. Curves show model-predicted crust thicknesses and dots represent actual data points. Limestone types are (CR) Cordova Cream; (CS) Cordova Shell; (LD) Lueders; and (BD) Bedford. *Source*: Bandyopadhyay, J. K., Annamalai, S., and Gauri, K. L., Application of artificial neural networks in modeling limestone–SO_2 reaction, *Am. Inst. Chem. Eng. J.*, **42**(8): 2295–2302, 1996. Reproduced with the permission of the American Institute of Chemical Engineers, copyright 1996 AIChE.

SUMMARY

We have modeled marble, limestone, and dolomite reactions using experimental rate data generated in concentrated atmospheres of SO_2 and NO_2. The prediction by model equations match weathering in outdoor conditions as confirmed by certain experiments performed outdoors. The aspects of weathering that can be predicted include crust thickness, surface recession, and porosity development.

REFERENCES

Bhatia, S. K., and Perlmutter, D. D., A random pore model for fluid-solid reactions: I. Isothermal, kinetic control, *Am. Inst. Chem. Eng. J.*, **26**(3): 379 (1980).

Bird, R. B., Stewart, W. E., and Lightfoot, E. N., *Transport Phenomena* New York: Wiley, 1960, p. 780.

Mazet, N., and Spinner, B., Modeling of gas-solid reactions. 2. Porous solids, *Int. Chem. Eng.*, **32**(3): 395–407, 1992.

Marquardt, D. W., An algorithm for least squares estimation of nonlinear parameters. *SIAM J. Appl. Mathematics*, **11**: 431–435, 1963.

Reddy, M. M., and Baedecker, P. A., Acidic deposition: State of science and technology, National Acid Precipitation Assessment Program, Report 19.

Szekely, J., Evans J. W., and Sohn, H. Y., *Gas–solid reactions*, 1st ed., New York: Academic, 1976.

Zarkanitis, S., and Sotirchos, S. V., Pore structure and particle size effects on limestone capacity for SO_2 removal, *Am. Inst. Chem. Eng. J.*, **35**(5): 821, 1989.

CHAPTER 7

Biodeterioration

7.1 INTRODUCTION

Biodeterioration has been described as a black, white, green, or red rock disease (Krumbein, 1988), presumably after the color of a surface film imparted by microbial encrustation, called a biofilm. In addition, certain organisms cause deterioration of the rock by biophysical and biochemical processes. Biophysical processes relate to the penetration of organelles into the cracks and pores, which split the rock. Biochemical breakdown can be assimilatory when organisms extract minerals (ions of calcium, iron, etc.) as nutrients, or dissimilatory when products of microbial metabolic activity, called metabolites, react with the rock and produce new minerals. If these minerals are insoluble in water they produce variably colored crusts; if soluble the stone is rendered friable as the minerals are leached. Also, microorganisms aid in the migration of iron, which is often found in carbonate rocks, toward the stone surface. Here, the iron undergoes a volume increase on oxidation and causes some disintegration, while also imparting hues of red, brown, and green color to the stone surface.

The organisms responsible for biodeterioration are called *biodeteriogens*. Although all types of organisms can break down rock, lower forms of life, particularly microflora, are considered the major cause of deterioration. We give here the organization of the microflora consisting of bacteria, mosses and liverworts, algae, fungi, and lichens and their effect as biodeteriogens. Later in this chapter we will describe the work of higher forms of life, such as vascular plants, in the deterioration of stone.

7.2 ORGANIZATION OF MICROFLORA

Microflora are categorized as autotrophs and heterotrophs by the mode of production of energy needed for their metabolic activity.

> *Autotrophs.* Autotrophs are organisms that can independently manufacture food. For example, *photoautotrophs* are mostly green plants. They contain chlorophyll, which enables them to combine carbon dioxide and water in the presence of sunlight. This process, called photosynthesis, produces carbohydrates:

$$CO_2 + H_2O + energy \rightarrow CH_2O + O_2$$

The energy stored in the carbohydrates (CH_2O) is later released by the organisms for their vital needs through fermentation or respiration reactions. In the fermentation reaction, carbohydrate molecules form carbon dioxide, an alcohol (ROH, where R is commonly methyl or ethyl radical), and energy. In respiration, carbohydrates combine with oxygen to produce carbon dioxide, water, and energy.

$$CH_2O \rightarrow CO_2 + ROH + e \quad \text{(fermentation)}$$

$$CH_2O + O_2 \rightarrow CO_2 + H_2O + e \quad \text{(respiration)}$$

Another form of autotroph, *chemoautotrophs*, lacks chlorophyll and therefore cannot rely upon solar energy to manufacture food. They "breathe" complex ions such as sulfates (SO_4^{2-}) and nitrates (NO_3^-) to produce energy.

Heterotrophs. Heterotrophs are organisms that get their energy by eating autotrophs or other heterotrophs. They are also organized as photoheterotrophs and chemoheterotrophs depending upon the substrate upon which they thrive.

Among organisms that break down rock, some are autotrophs, others heterotrophs, and some are even mixotrophs, that is, they change their behavior in response to change in ecological conditions. Even within each group, for example, bacteria, some are autotrophs other heterotrophs.

Another factor to consider in the classification is whether the organisms are aerobic or anaerobic. Aerobic organisms are those that can exist in an oxygenated atmosphere while anaerobic organisms can survive only in an oxygen-devoid atmosphere. Based upon these factors, Table 7.1 gives the classification of the lower form of organisms involved in the deterioration of rocks.

7.3 DESCRIPTION AND EFFECTS OF MICROFLORAL BIODETEROGENS

7.3.1 Bacteria

Photoautotrophs. Common photoautotrophic bacteria are the *Cyanobacteria*, often called blue-green algae. They appear as slimy coatings on stone surfaces. *Cyanobacteria* are characterized by the presence of chlorophyll and other pigments in their cells. Because the pigments come in nearly all colors of the visible light, they can impart corresponding colors to the rock upon which they grow. Thus, their main contribution to biodeterioration is the discoloration of the stone surface. Some believe that these biofilms are decorative and even protect the underlying masonry. However, certain *Cyanobacteria* can fix

TABLE 7.1 Classification of Microflora Involved in the Deterioration of Carbonate Rocks

Nutritional Category	Energy Source	Carbon Source	Electron Donors	Electron Acceptors	Groups of Organisms
Photoautotrophs or photolithotrophs	Sunlight (photosynthetic organisms)	CO_2	Water	Oxygen	Aerobic organisms: bacteria (*Cyanobacteria*), algae *Bacillariophyta* or diatoms), mosses and liverworts, higher plants, algae (*Cholorophyta*)
			H_2S, H_2	Organics	Anaerobic organisms: Green and purple sulfur bacteria, purple nonsulfur bacteria
Chemoautotrophs or chemolithotrophs	Redox reaction (chemosynthetic organisms)	CO_2	H_2, Fe^{2+}, NH_4^+, NO_2^-, S, SO_3^{2-}	Oxygen	Aerobic organisms: Hydrogen bacteria, iron bacteria, nitrifying and nitrogen-fixing bacteria, sulfur oxidizing bacteria
			S, SO_3^{2-}, H_2S	NO_3^-;	Anaerobic organisms: Sulfur-reducing bacteria
Photoheterotrophs or photoorganotrophs	Sunlight (photosynthetic organisms)	Organics	Organics	Oxygen	Aerobic organisms: Eukaryotic algae, lichens
Chemoheterotrophs or chemoorganotrophs	Redox reactions (chemosynthetic organisms)	Organics	Organics	Oxygen Organics, NO_3^-, SO_4^{2-}, SO_3^{2-}, $S_2O_3^{2-}$	Aerobic organisms: Animals Anaerobic organisms: Fermentative bacteria, fungi, actinomycetes, sulfur-reducing bacteria

Source: Personal communication, to be published by Kumar, R., and Kumar, A.V., *Biodeterioration of Stone in Tropical Regions*, Los Angeles, The Getty Conservation Institute.

atmospheric nitrogen as ammonium sulfate, which is then able to decay stone when oxidized to sulfuric acid by sulfur-oxidizing bacteria (discussed later).

Chemoautotrophs. Two classes of chemoautotrophic bacteria are considered common biodeterogens. They are sulfur bacteria and nitrifying bacteria. Sulfur bacteria metabolize sulfur compounds present in the rock and the nitrifying bacteria fix atmospheric nitrogen as well as utilize nitrogen-bearing compounds present in the rock. Among each of these classes are certain bacteria, which either oxidize or reduce the sulfur and nitrogen compounds.

- *Sulfur-Oxidizing Bacteria.* Thiobacilli are common sulfur-oxidizing bacteria. Certain *Thiobacillus* species are not acid tolerant. Therefore, they oxidize reduced sulfur compounds, such as hydrogen sulfide (H_2S), to elemental sulfur (S). Other *Thiobacillus* species are acidophillic. They oxidize the reduced sulfur to sulfuric acid according to the reaction

$$S + 1\tfrac{1}{2}O_2 + H_2O \rightarrow H_2SO_4$$

This and other sulfur-oxidation reactions are exothermic reactions. The energy generated in these reactions is then used by *Thiobacilii* for metabolism.

- *Sulfur-reducing Bacteria.* Geologically important sulfur-reducing bacteria belonging to the species *Desulphovibrio desulphuricans* are discussed in the chapter on conservation. These bacteria have the capacity to convert calcium sulfate to calcium carbonate. However, in the process elemental sulfur is produced, which may be converted to sulfuric acid by the sulfur-oxidizing bacteria.

- *Nitrifying Bacteria.* Nitrifying bacteria oxidize nitrogen compounds such as ammonia. The source of ammonia is often the atmospheric nitrogen converted by nitrogen-fixing bacteria commonly found in soils. However, several nitrifying bacteria have been isolated from rock, each oxidizing the nitrogen compound to different level. For example, *Nitrosomonas* convert ammonia to nitrite and then *Nitrobacter* convert nitrite to nitrate, according to the reactions

$$NH_4 + 1\tfrac{1}{2}O_2 + H_2O \rightarrow NO_2 + 2H_2 + H_2O$$

$$NO_2 + \tfrac{1}{2}O_2 \rightarrow NO_3$$

The nitrite and nitrate convert to nitrous and nitric acid in the presence of water, which then corrode the rock.

May and Lewis (1988, p. 60) state that, "Nitrogen and sulfur bacteria are thus capable, by their biologic activity, of generating the same acids which may be present in acid rain." Figure 7.1 is adapted from their study, showing correlation between the reaction products of bacterial activity and those produced by atmospheric gases on carbonate rocks.

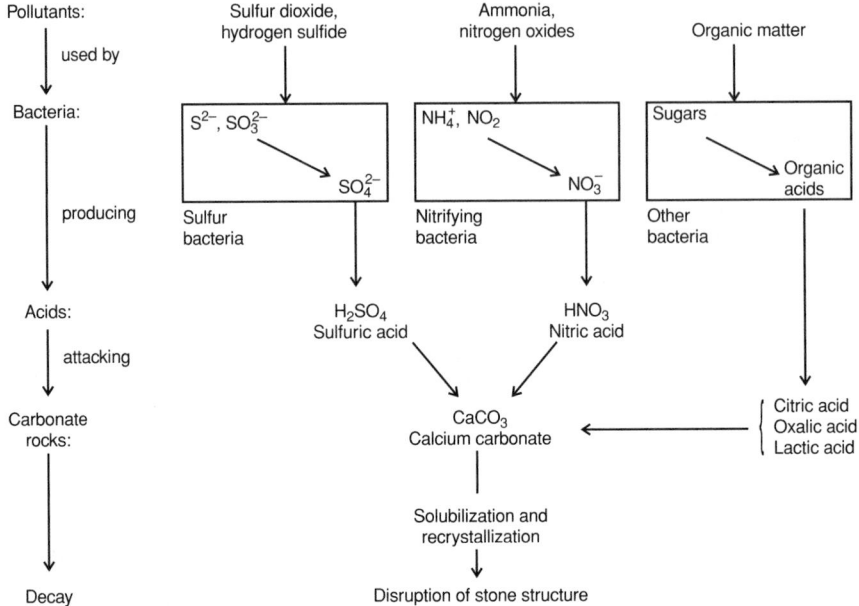

FIGURE 7.1 Possible involvement of bacteria in carbonate rock decay. Correlation of chemical reactions due to atmospheric pollutants and biochemical reactions by bacteria is shown. *Source*: May, E. and Lewis, F. J., Strategies and techniques for the study of bacterial populations on decaying stonework, in *The VIth International Congress on Deterioration and Conservation of Stone*, J. Ciabach (ed.), Torun: Nicholas Copernicus University Press, 1988, pp. 59–70.

7.3.2 Lichens

Lichens are symbiotic associations of algae and fungi. While algae are photoautotrophic, the fungi are chemoheterotrophic organisms. The association of the two provides for survival in the most drastic conditions. In favorable conditions of humidity and temperature, algae thrive, building up the nutrients that then provide subsistence for fungi when adverse conditions arrive. Therefore, lichens are believed to be the first in the food chain, inhabiting new terrain and creating conditions favorable for more sensitive communities to arrive later. Certain lichens, called calcicolous, colonize carbonate rocks while silica-bearing rocks, such as sandstone and igneous rocks, are colonized by silicolous lichens.

Algae are largely responsible for the formation of multicolored mats on the stone surface, changing the appearance but not the structural integrity of the stone. Fungi, on the other hand, penetrate the rock mechanically and also produce organic acids. The overall deterioration effect of lichens is thus fine powdering of the stone surface by the penetration of rhizoids (rootlike filaments) and hyphae (fruiting bodies) and a slow corrosion of calcite and dolomite by organic acids and chelating complexes (Fig. 7.2).

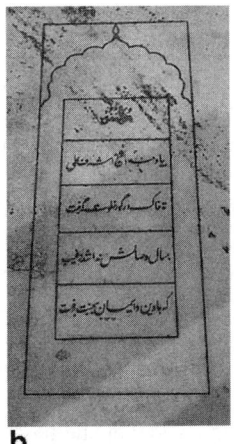

a   b

FIGURE 7.2   Corrosion of marble by lichens. Pitting by lichens of a marble tombstone from Fatehpur Sikri, India, can be seen. Linear pitting, shown in (b) is due to oriented marble fabric.

Oxalic acid (H_2CO_4) is one of the metabolites of lichens that reacts with the carbonate rocks, forming a crust made of the minerals wedelite and whewellite ($CaC_2O_4 \cdot H_2O$). Such a crust is considered by some as being protective. Nevertheless, cases are known in which the lichens have proved to be protective without the formation of oxalate crust. For example, Schaffer (1972) states that "blocks of stone quarried at the time of erection of St. Paul's Cathedral, and left lying in the Isle of Portland for 150 years, still exhibited the tool marks below the lichens growth. Similarly, the tool marks are still clearly visible beneath lichens growths on the masonry of the Wells Cathedral and no appreciable damage can be detected."

Finally, the effects of microbial biodeterogens can be summarized as follows:

- Encrustation, which may be protective.
- Acidolysis, which is defined as the dissolution of carbonate minerals in inorganic and organic acids. Inorganic acids are strong; they cause removal of a large amount of carbonate rock. Organic acids are rather weak compared with inorganic acids; they cause mere pitting of the rock.
- Chelation, the removal of cations, such as calcium, from calcite by bonding with organic acids. Chelation is a rather slow process and results in fine powdering or pitting of the rock.

7.4   STONE DETERIORATION BY HIGHER PLANTS

Higher plants are those that have a vascular system to distribute minerals throughout the plant. These plants derive nutrients from soil or bare rock through a well-developed root system. Roots break down rocks by the chemical reactions and mechanical forces discussed in the following.

The chemical effects of roots include acidolysis, chelation, and ion exchange. Rhizosphere is the interface between the root and the soil (Fig. 7.3). It is enriched in hydrogen ions due to the pH of the root surface, which is often below 6. Furthermore, roots produce carbon dioxide by respiration, which in the presence of water forms carbonic acid. Furthermore, roots produce organic acid. Outside the rhizosphere, cations of Ca^{2+} and Mg^{2+} are present in carbonate rock in greater abundance if clay minerals are also present. These cations exchange with the H^+ ion of rhizosphere. By this exchange as well as by reaction with carbonic and organic acids, the carbonate rock is dissolved.

Mechanical disintegration of the rock occurs by the physical force of the growing root. Often this force is less than the penetration resistance of the bedrock. It is believed that chemical breakdown first reduces the penetration resistance and then the growing tip of the root is able to open and widen cracks in the rock.

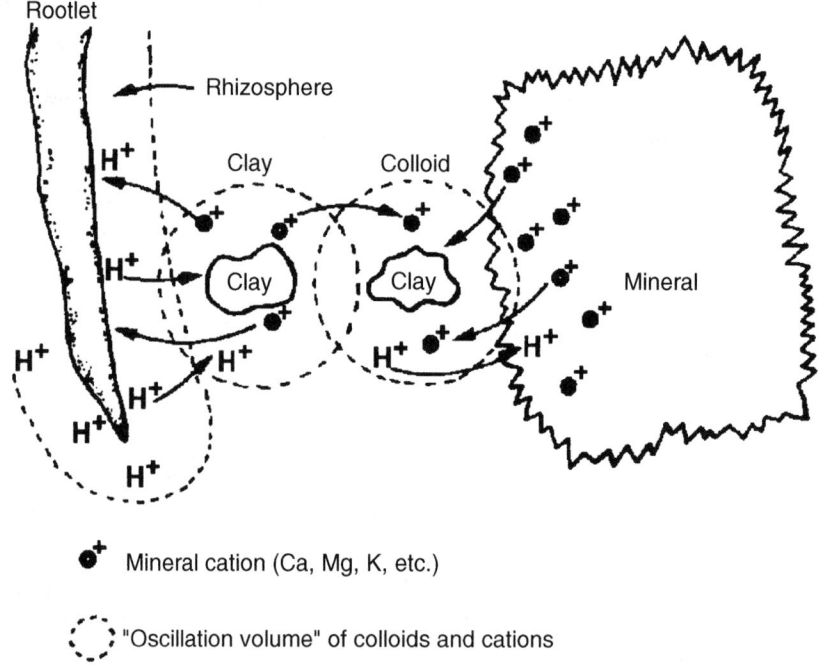

FIGURE 7.3 Chemical corrosion by roots of plants. This shows the acidic environment of the rhizosphere and its effect upon the surrounding minerals of the bedrock. Only cation exchange is shown, but acidolysis and chelation are also at work. *Source*: Caneva, G., Biochemical mechanisms of stone weathering induced by plant growth, in *The VIth International Congress on Deterioration and Conservation of Stone*, J. Ciabach (ed.), Torun: Nicholas Copernicus University Press, 1988, pp. 32–44.

7.5 OVERVIEW OF BIODEGRADATION BY MICROBES

The expressions of biodegradation by microbes are similar to those of physical and chemical weathering, namely, removal or loosening of surface material or formation of crust. A commonly cited proof of degradation by microbes is the association of bacteria with the weathered rock and the association of oxalate crust with lichens. Since bacteria have also been isolated from sound rock and at least some of the oxalate crusts are known to be coatings applied by restorers, some believe that biodegradation is actually a noninvasive phenomenon. In this view, the processes of physical and chemical weathering prepare a suitable substrate that microbes can easily colonize. Nevertheless, the action of lichens as biodetereogens is certain: Figure 7.3 shows marble in an environment that is practically devoid of air pollution yet is deeply pitted.

SUMMARY

Biodeterioration includes the formation of biofilms and biophysical and biochemical changes in the rock. Biofilms impart various colors to the stone surface, which may be considered by some as discoloration; others view it as enhancement of stone appearance.

Biophysical disintegration occurs by penetration of organelles of microbes and roots of vascular plants into stone crevasses. Biochemical decay involves extraction of nutrients, exchange of cations, and chelation and solubilization of minerals. Certain biochemical processes result in the formation of new minerals, which produce crust upon the stone surface. An oxalate crust formed in this fashion is considered protective.

REFERENCES

Krumbein, W. E., Biology of stone and minerals in buildings—biodeterioration, biotransfer, bioprotection, in *The VIth International Congress on Deterioration and Conservation of Stone*, J. Ciabach (ed.), Torun: Nicholas Copernicus University Press, 1988, pp. 1–13.

May, E. and Lewis, F. J., Strategies and techniques for the study of bacterial populations on decaying stonework, in *The VIth International Congress on Deterioration and Conservation of Stone*, J. Ciabach (ed.), Torun: Nicholas Copernicus University Press, 1988, pp. 59–70.

Schaffer, R. J., *The Weathering of Natural Building Stones*, British Department of Scientific and Industrial Research, Building Research, Special Report No. 18, p. 74, reprinted, 1972.

CHAPTER 8

Methods of Characterization of Limestone and Dolostone by Mercury Porosimetry

8.1 INTRODUCTION

Mercury porosimetry is one of the methods to determine pore size and pore volume. We have shown how weathering of porous carbonate rocks is dependent on the presence of pores in the stone. We will give later in Chapter 9 the application of pore-size distributions to derive durability factors of limestone. Thus to quantify the process of weathering it is necessary to determine pore properties. Mercury porosimetry is a convenient, a quantitative, and the most efficient method for determining sizes and volumes of pores in a solid material.

Mercury porosimetry consists of injecting mercury into pores under pressure. The size of pores intruded is inversely proportional to the applied pressure. The Washburn equation is commonly used to determine pore radii. In addition to pore sizes, porosity, pore-surface area, and density can also be determined from the porosimetric data. Furthermore, the pressure–volume (PV) data can be used in terms of thermodynamic work and to express the fractal structure of pores. This chapter describes methods to determine quantitatively the various pore properties of limestone and dolomite and to characterize these by pore-size distributions, pore potential, and fractal dimension.

8.2 PRINCIPLE AND METHOD OF MERCURY POROSIMETRY

The principle of mercury porosimetry is that mercury, a nonwetting, nonreactive liquid, can be intruded into pores by isostatically increasing the pressure and that the pore size intruded is inversely related to the pressure. Thus, the pore radius and associated volume can be obtained from the equilibrium pressure. Additionally, the pore-surface area, total pore volume, and bulk and skeletal densities can also be determined.

The instrument used to measure pore size by mercury intrusion is called a pore-sizer or porosimeter (Fig. 8.1). A porosimeter consists of a low- and a high-pressure port. Often, a porosimeter is a fully automated system run by commands

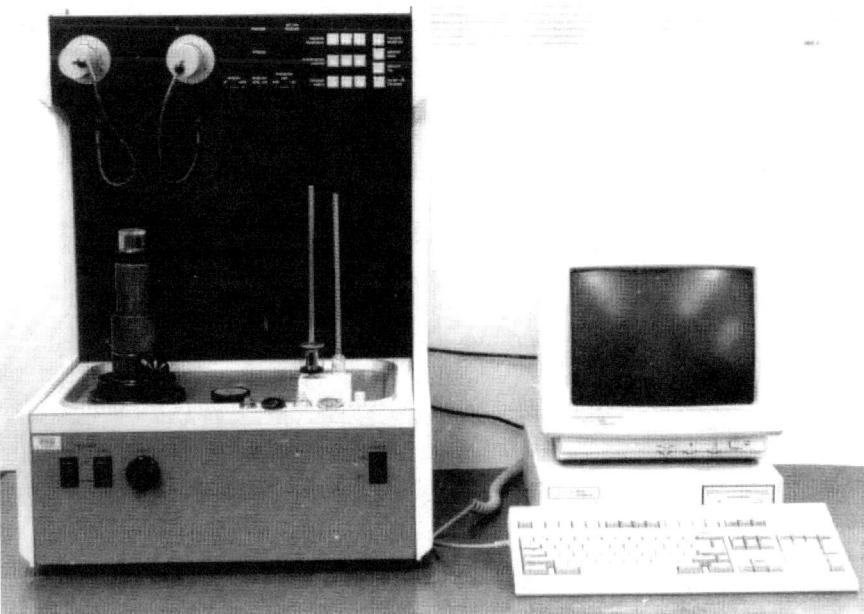

FIGURE 8.1 Pore-sizer. The 9320 *Micromeritics* Pore Sizer is shown. Two low-pressure ports (white) are on the upper left of the panel and the high-pressure chamber is the column beneath these. Two penetrometers are seen as vertical rods on the lower right in front of the panel. This is a fully automated system run by commands given at the computer.

given at the computer. Data collection, data reduction, data display, and all aspects of high-pressure analysis are performed by control modules. The data analysis is represented as a printout in tabular and graphic forms.

The sample, commonly a core from the rock, is often cleaned in an ultrasonic bath, dried at 105°C, cooled in a desiccator, and weighed for density measurement. The sample is then placed in a penetrometer (sample cell), which is selected on the basis of sample size and the expected pore volume. The penetrometer selection guide is often given in the manufacturer's manual.

In the system shown in Fig. 8.1, low-pressure readings are obtained in the subatmospheric pressure in the range of nearly 5 psia to 15 psia. For these readings, a vacuum pump is used to evacuate the sample. After the sample is pumped down, the penetrometer is filled with mercury introduced through a manifold, and then air is bled to increase the manifold pressure, in steps if several low-pressure readings are desired. An auxiliary pressure source, for example, compressed air, can be used to increase the pressure somewhat above the atmospheric pressure. After the low-pressure readings are taken, the penetrometer is transferred to the high-pressure port.

High-pressure readings are obtained by increasing the pressure generated by a hydraulic pump. High pressure can be raised to 60,000 psia in most instruments; however, pressure up to 30,000 psia is often adequate to fill most pores in carbonate rocks.

In the low- as well as high-pressure range, the volume of intruded mercury is measured in terms of capacitance as picofarads and converted to cm^3/g of sample using conversion factors given by the pore-sizer manufacturer.

8.3 POROSITY, DENSITY, AND PORE-SURFACE AREA

From the PV data of the high-pressure range (sample data are shown in Table 8.1) and from certain weights recorded before and after low-pressure intrusion, the bulk and skeletal porosity, density, and the pore-surface area can be calculated. The relations to calculate bulk and skeletal densities and the pore-surface area are adapted from Anon. (n.d.)

Porosity ε, sometimes called bulk porosity, is the ratio of total pore volume (p_v) to the volume of the sample V_s, given as

$$\varepsilon = \frac{p_v}{V_s} \tag{8.1a}$$

or

$$\varepsilon = 1 - \frac{\rho_b}{\rho_s} \tag{8.1b}$$

TABLE 8.1   PV Data for Initial Intrusion, Extrusion, and Reintrusion and Calculated (dV and Av. P) Values Used for Representing Cumulative and Incremental (Differential) Intrusion (Figs. 8.2 to 8.4)

Pressure (PSI)	Intrusion (cc/gm)	Extrusion (cc/gm)	Reintrusion (cc/gm)	dV Intrusion	dV Extrusion	dV Reintrusion	Av. P (psi)
30	0.008	0.079	0.076	—	—	—	—
50	0.024	0.082	0.077	0.016	0.003	0.001	40
100	0.0565	0.0853	0.0805	0.0325	0.0033	0.0035	65
200	0.06909	0.0865	0.084	0.01259	0.0012	0.0035	150
300	0.075	0.087	0.086	0.00591	0.0005	0.002	250
500	0.078	0.088	0.0878	0.003	0.001	0.0018	400
700	0.08	0.0882	0.0884	0.002	0.0002	0.0006	600
1000	0.081	0.0885	0.0885	0.001	0.0003	1E-04	850
2000	0.083	0.0895	0.0895	0.002	0.001	0.001	1500
4000	0.086	0.091	0.091	0.003	0.0015	0.0015	3000
6000	0.088	0.092	0.092	0.002	0.001	0.001	5000
8000	0.0895	0.0928	0.0928	0.0015	0.0008	0.0008	7000
10000	0.091	0.0935	0.0935	0.0015	0.0007	0.0007	9000
20000	0.0953	0.0965	0.0965	0.0043	0.003	0.003	15000
30000	0.0985	0.0985	0.0985	0.0032	0.002	0.002	25000

where ρ_b and ρ_s are bulk density and skeletal density, respectively. While the pore volume is given by the total volume of mercury intruded, the volume of the sample V_s can be obtained as follows:

$$V_s = V_{pen} - V_{Hg} \tag{8.2}$$

where V_{pen} is the total internal volume of the penetrometer and V_{Hg} is the volume of mercury occupying space between sample and penetrometer.

To determine V_{Hg} a number of weights are taken using an analytical balance. These weights are

- W_s: weight of dry sample
- W_1: weight of the sealed penetrometer with only the sample inside
- W_2: weight of the sealed penetrometer with the sample inside and filled with mercury

Thus, the volume of mercury occupying the remaining space around the sample is

$$V_{Hg} = \frac{W_2 - W_1}{\rho_{Hg}} \tag{8.3}$$

where ρ_{Hg}, the density of mercury at 25°C, is 13.5335. Now,

$$V_s = V_{pen} - \frac{W_2 - W_1}{\rho_{Hg}} \tag{8.4}$$

Substituting V_s and p_v [mercury intruded at highest pressure (Table 8.1)] in Eq. (8.1a), gives the porosity of the sample.

The *bulk density* (ρ_b) is the weight per unit volume (g/cm^3) of the sample. Density can be determined from the dry weight, W_s(g) of the sample, taken before it was placed in the penetrometer, and its volume determined by the application of Eq. (8.4), as

$$\rho_b = \frac{W_s}{V_s} \tag{8.5}$$

The *skeletal density* (ρ_s) is the ratio of weight of the sample to its volume (g/cm^3) without pores:

$$\rho_s = \frac{W_s}{V_s - p_v} \tag{8.6}$$

The *pore-surface area* (P_A) can be calculated in two ways: (1) by considering the PV data as work and (2) by considering the cylindrical pore geometry.

The basic control over mercury penetration is the resisting surface tension (Chap. 4.3, Fig. 4.8), which is related to pore radius and pore volume as

$$\gamma \cos \theta \, dA = - P dV \tag{8.7}$$

or

$$A = - \frac{PV}{\gamma \cos \theta} \tag{8.8}$$

where γ is the surface tension of mercury, 485 dyn/cm, θ is the contact angle of mercury with the pore wall, 130°, and V is the volume of intruded mercury at applied pressure P. Equation (8.8) for the evaluation of the pressure–volume mercury penetration data becomes

$$A = - \frac{\sum P \Delta V}{\gamma \cos \theta} \tag{8.9}$$

Regarding the calculation of pore-surface area from cylindrical pore geometry, the relationship among pore area, radius, and volume for a cylinder is [see Eq. (8.13)]

$$A = \frac{2V}{r} \tag{8.10}$$

$$r = - \frac{2\gamma \cos \theta}{P}$$

Thus

$$A = - \frac{PV}{\gamma \cos \theta} \tag{8.11}$$

which, as before, when written for evaluation from pressure–volume mercury penetration data, becomes

$$A = - \frac{\sum P \Delta V}{\gamma \cos \theta} \tag{8.12}$$

In the following we will give sample calculations for determining these properties for the sample for which PV data is shown in Table 8.1. Pertinent properties of that sample and other related parameters are $W_s = 3.6685$ g; $W_1 = 68.3027$ g; $W_2 = 145.0913$ g; $V_{pen} = 7.5$ cm^3; pore volume = 0.0985 cm^3 g; total pore volume, $p_V = 0.3613$ (3.6685 × 0.0985) cm^3,

$$V_{Hg} = \frac{145.0913 - 68.3027}{13.5335} = 5.674 \, cm^3$$

by Eq. (8.3), and $V_s = 7.5 - 5.674 \, cm^3 = 1.826 \, cm^3$ by Eq. (8.4). Thus, the porosity $\varepsilon = 0.3613/1.826 = 0.1978 \cong 19.8\%$, by Eq. (8.1). The bulk density (ρ_b) and skeletal density (ρ_s) are $\rho_b = 3.6685/1.826 = 2.009 \, g/cm^3$, by Eq. (8.5) and $\rho_s = 3.6685/(1.826 - 0.3613) = 2.5046 \, g/cm^3$ by Eq. (8.6). The total porosity can also be obtained from the bulk and skeletal densities. $\varepsilon = 1 - 2.009/2.5046 = 0.1978 \cong 19.8\%$, by Eq. (8.1b).

Now, we can calculate the pore-surface area. From the PV data, where P is in the range of 5 psia to 30,000 psia, and V is the volume in cm^3/g of sample, knowing $\theta = 130°$ and $\gamma = 485 \, dyn/cm$ and using Eq. (8.12), we get

$$A = \frac{-\sum P \Delta V}{\gamma \cos \theta} \qquad (8.13)$$

$$A = \frac{-1.37362 \times 10^7}{485 \cos(130°)} \frac{(dyn/cm^2)(cm^3/g)}{dyn/cm}$$

$$A = 4.406 \, m^2/g$$

Thus, the porosimetric data allow determination of some basic properties of porous media. In addition pore-size distributions are obtained, which we will describe later in detail. At our ftp site (see Appendix C), we provide programs to determine these properties.

Porosimetric data consist of pressure and associated volume of mercury for many recordings. Sometimes it is necessary to obtain PV values for points between those recorded by the pore-sizer. This is done by commonly available interpolation techniques such as one based on the spline algorithm.

The pressure–volume data can be viewed in a variety of ways, including (1) pore-size distributions, (2) fractal pore structure, and (3) thermodynamic work. Since the PV data are unique to each stone type, all these methods can be used to characterize any porous medium. We will discuss these methods now and use these to characterize limestone for durability in the following chapter.

8.4 PORE-SIZE DISTRIBUTIONS

Pore size is calculated from the pore-volume data by the application of the Washburn equation, given as

$$r = -\frac{2\gamma \cos \theta}{P} \qquad (8.14)$$

where r is the radius (in μm $= 10^{-3}$ mm), γ is the surface tension of mercury, 485 dyn/cm, θ is the contact angle of mercury with the pore wall, 130°, and P is the applied pressure (psia). Thus

$$r = \frac{90}{P}$$

In the application of porosimetry, the pressure commonly used to intrude mercury is in the range of 20 psia to 30,000 psia (Fig. 8.2). Consequently, volumes of pores with radii nearly 0.003 μm to 5 μm should be filled. However, in this pressure range large pores, defined here as being pores greater than 5 μm, connected to small pores are also filled. Thus, from the PV data that we have an erroneous view is obtained that the total volume of intruded mercury (Fig. 8.2) in the initial intrusion is the volume of small pores only. We will describe a method below whereby the volume of the large pores, greater than 5 μm, can be discerned from this data.

The porosimeter allows that, after the initial intrusion or at any level within the range of maximum pressure, the pressure can be reduced whereby the mercury is extruded from the pores. Experiments with connected small and large glass capillaries, where the movement of mercury can be seen, show that, if the pressure is reduced, mercury empties from the small capillaries only. This is explained by the "snap-off effect." According to this effect, the mercury in large capillaries becomes dissociated from the outside mercury immediately after the small capillary is emptied and remains trapped. Thus the volume of trapped mercury gives the volume of the large pores. For example, in Figure 8.3, the initial intrusion volume was 14×10^{-2} cm³/g and only $7 \times$

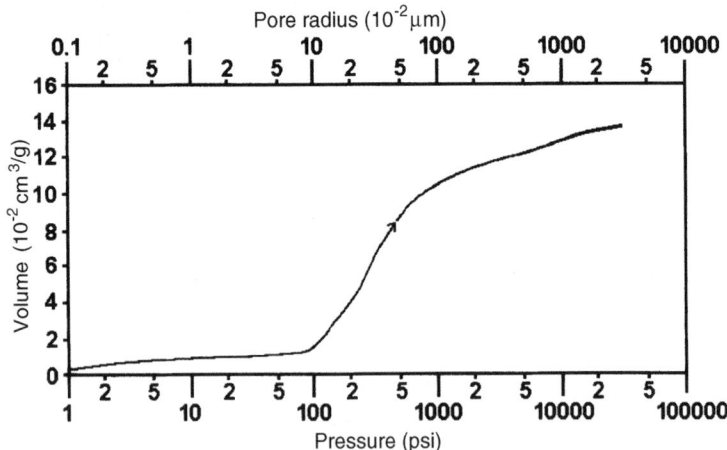

FIGURE 8.2 Cumulative intrusion versus pressure in initial intrusion. Data obtained in an initial intrusion up to 30,000 psia are shown as an incremental plot in which the volume of the intruded mercury at a given pressure is added to the volume in the preceding intrusion.

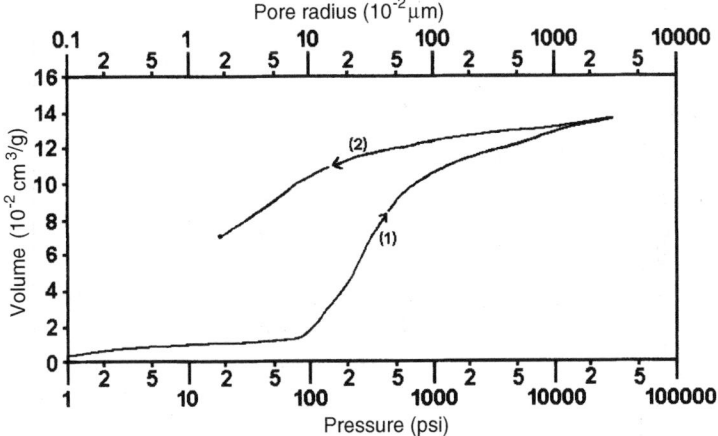

FIGURE 8.3 Cumulative initial intrusion and extrusion versus pressure. Data obtained in the initial intrusion (curve 1) and extrusion (curve 2) are shown. The amount of trapped mercury (difference between the initial intrusion and extrusion) indicates the volume of large pores, which is about 50% in this case.

10^{-2} cm^3/g was extruded, showing that large pores constitute 50% of the total pore volume.

Now, we can confirm the volume of the small pores if we intrude the mercury again (reintrusion) by increasing the pressure. It follows from the preceding discussion that in the reintrusion, only the small pores will be filled because the large pores were not emptied in the extrusion process. Figure 8.4(a) shows that this is true, even though some hysteresis in the extrusion and reintrusion is observed, which we will discuss later.

8.4.1 Cumulative and Incremental Distributions

The pressure–volume (*PV*) data can be represented in a variety of ways to show pore-size distributions. We will describe two methods that are commonly used:

- *Cumulative Intrusion Volume versus Pressure or Radii [Figs. 8.2 to 8.4(a)].* The cumulative intrusion volume at any pressure reading, or equivalent pore radius, is the sum of mercury intruded in all previous readings. When pressure is shown, sometimes its logarithmic values are given because the recorded values of pressure are high compared with the small volume of intruded mercury.

- *Incremental Intrusion Volume versus Pressure or Radii.* An incremental intrusion plot [Fig. 8.4(b)] presents the data in a differential form so that the intrusion volume between two pressure readings can be shown. The incremental intrusion plot may be made to show the intrusion volume as (1) $dV/\overline{P}$ versus P,

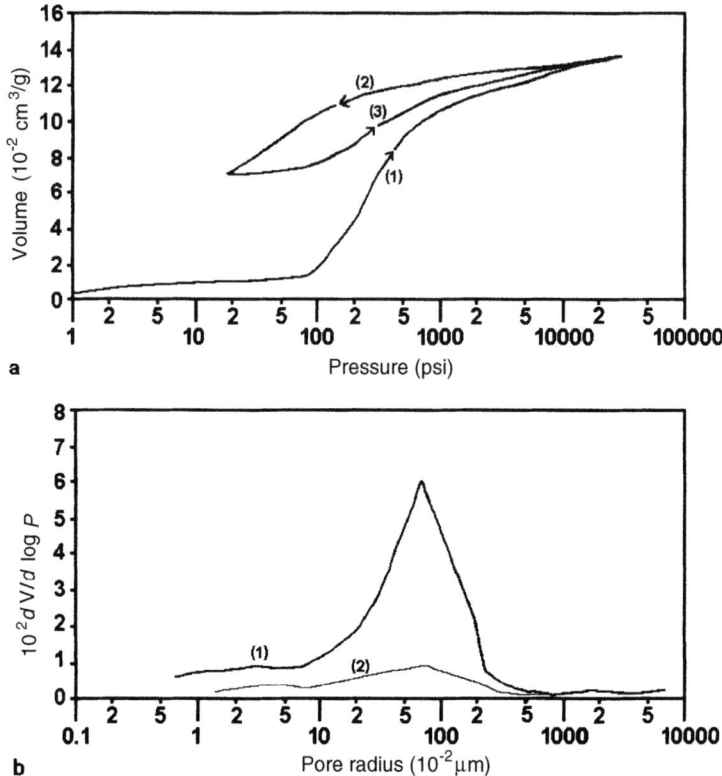

FIGURE 8.4 Cumulative and differential intrusion. (a) shows a cycle of cumulative initial intrusion, extrusion, and reintrusion. (b) shows (1) differential initial intrusion and (2) reintrusion curves. In differential intrusion the volume between two pressure readings is divided by the mean of these pressures. Notice that the initial intrusion volume is much larger than that in reintrusion because the mercury remained trapped in large pores which were intruded at the breakthrough pressure of the small pores. Notice also the maximum volume of small pores of nearly 0.5 μm radii to which the large pores are connected. *Source*: Punuru, A. R., Chowdhury, A. N., Kulshreshtha, N. P., and Gauri, K. L., Control of porosity on durability of limestone at the Great Sphinx, Egypt, *Environ. Geol. Water Sci.*, **15**(3): 225–232, 1990. Reproduced by the permission of Springer-Verlag. Copyright 1990 Springer-Verlag.

where dV and $\overline{P}$ are the difference of volume and the mean pressure between the two pressure readings, and (2) $dV/d \log P$ versus P [Fig. 4(b)], where $d \log P$ is the difference between the logarithm of pressures at two readings, $d \log P$ being used to normalize the data. This form of plot immediately shows modal pore distribution by peaks, for example, the unimodal distribution is shown by a single peak [Fig. 8.4(b)] and the bimodal distribution, if present, by two peaks. Table 8.1 gives PV data for a limestone sample and calculated values of pressure and volume that may be used for representing data by these methods.

8.5 PORE POTENTIAL

PV data can also be treated as thermodynamic work, which can be expressed as the pore potential, defined as the difference in energy, or work, required between injecting and extruding mercury from pores. The model of the pore potential is based upon the fundamental concept that work is performed on injection of mercury into pores under pressure and that the volume of injected mercury is proportional to the work. The work can thus be described as a function:

$$W = \int P dV \qquad (8.15)$$

We have noted in the foregoing that two hystereses occur in the process of initial intrusion, extrusion, and reintrusion shown in Figures 8.5 and 8.6. While the hysteresis between initial intrusion and extrusion is due to the entrapment of mercury in large pores, the hysteresis between extrusion and reintrusion indicates that greater pressure is needed to reintrude than to extrude the same volume of mercury. A coincident hysteresis loop is developed over and over again if the operation of reextrusion and reintrusion is repeatedly performed. This hystersis can be termed as the hystersis loop. It is evident that this loop relates to the small pores (throats) connecting the large pores (bellies) and is a result of not fully understood complex connections among the small pores. We will now develop a mathematical model for pore potentials to characterize different stone types.

The work associated with initial intrusion and extrusion can be expressed as

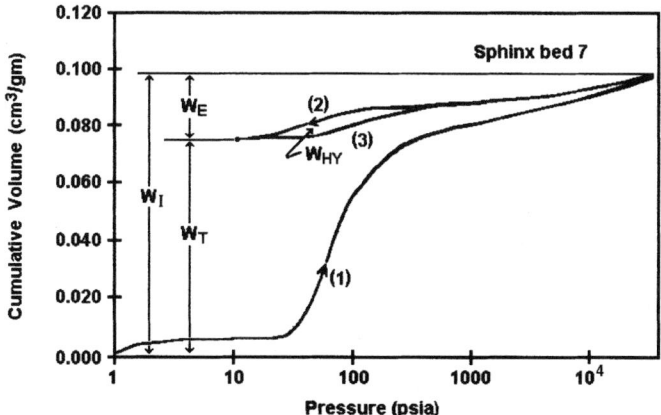

FIGURE 8.5 Work of hysteresis in intrusion and extrusion. W_I is the work of initial intrusion, W_T the work of entrapment, and W_E the work of extrusion. W_{HY} represents the work in the hysteresis loops between extrusion and reintrusion. *Source*: Gauri, K. L. and Yerrapragada, S. S., Pore potential and durability of Sphinx limestone, in *Materials Issues in Art and Archaeology III*, Vandiver, P. B., Druzik, J. R., Wheeler, G. S., and Freestone, I. C. (eds.), Symposium Proceedings, Pittsburgh: Materials Research Society, 1992, p. 918.

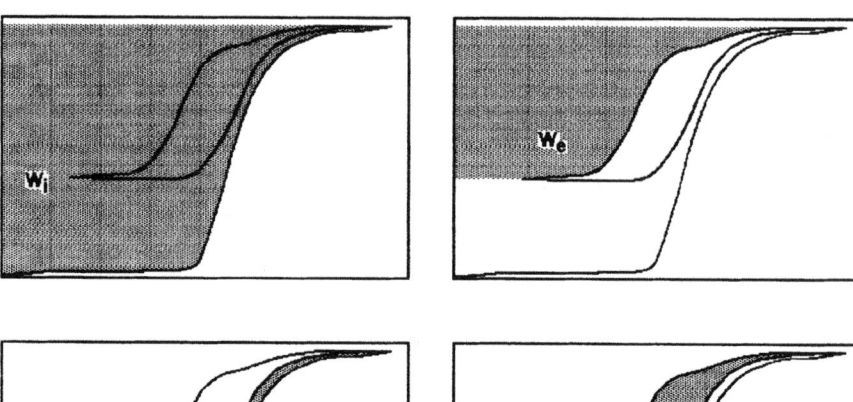

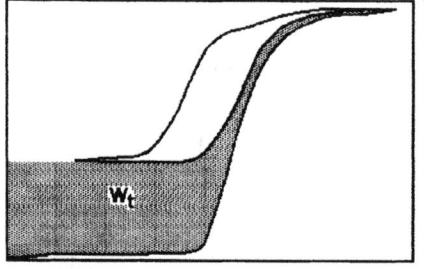

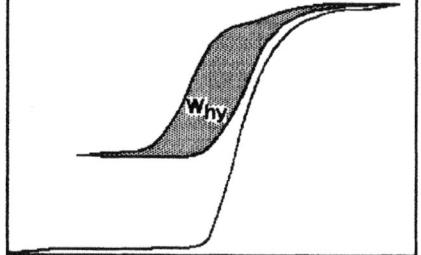

FIGURE 8.6 Work of hysteresis in intrusion and extrusion. This is the same as Figure 8.5, but the areas of work in different hysteresis are more clearly delineated.

$$\int dW_1 = \int P_I dV_I + \int P_E dV_E > 0 \qquad (8.16)$$

where P_I and P_E are the initial intrusion and extrusion pressures and V_I and V_E are the corresponding intrusion and extrusion volumes. Similarly, the PV work for reintrusion and extrusion cycle can be expressed as

$$\int dW_2 = \int P_{RE} dV_{RE} + \int P_E dV_E > 0 \qquad (8.17)$$

where P_{RE} and V_{RE} are the reintrusion pressures and volumes.

The PV work, for n radii intervals, can be calculated by numerical summation as

$$W = \sum_{}^{n} P\Delta V \qquad (8.18)$$

Looking at Figures 8.5 and 8.6, we note

$$W_I = W_T + W_{HY} + W_E \qquad (8.19)$$

where W_I is the PdV work for initial intrusion, W_T is the work for entrapment in large pores, W_{HY} is the work in the hysteresis loop between extrusion and reintrusion, and W_E is work for extrusion. These works are shown graphically in Figure 8.5 and 8.6. The PV work difference between intrusion and extrusion can be explained by means of a pore potential U, (Lowell and Shields, 1981) given as

$$U = \int F_I dL_I - \int F_E dL_E \tag{8.20}$$

where F_I and F_E are the intrusion and extrusion forces and L_I and L_E are the lengths of the mercury column. Alternately, the pore potential can be given as

$$U = \int P_I dV_I - \int P_E dV_E = W_I - W_E \tag{8.21}$$

Using Eq. (8.19), Eq. (8.21) can be written as

$$U = W_I - W_E = W_T + W_{HY} \tag{8.22}$$

Table 8.2 shows these values.

Now, we give a sample calculation to convert PV data to thermodynamic work (in joules per gram or J/g) for the pressure range of 20,000 psia to 30,000 psia (Table 8.2):

$$P\Delta V = 25,000 \text{ psi } (\overline{P}) \times 0.0032 \text{ cm}^3/\text{g } (\Delta V)$$

$$= 25,000 \times 6.8947 \times 10^4 \text{ dyn/cm}^2 \times 0.0032 \text{ cm}^3/\text{g}$$

$$= 5.5176 \times 10^6 \text{ dyn} \cdot \text{cm/g}$$

$$= 5.5176 \times 10^6 \times 10^{-7} = 0.55176 \text{ J/g}$$

Thus, for the entire range of the porosimetric cycle shown in Table 8.2, $U = 1.3736 - 0.8218103 = 0.5518$ J/g, by Eq. (8.22). Our ftp site (see Appendix C) contains programs to calculate the pore potential.

The application of the pore potential to characterize porous stone seems to offer a superior method than that by pore-size distributions because the calculation of the pore potential circumvents the assumption of cylindrical pores that is needed for the calculation of pore-size distributions.

8.6 FRACTAL DIMENSION

Another method by which porous material can be characterized is the fractal structure. Euclidean geometry classifies objects into linear, planar, or volumetric forms, which

TABLE 8.2 PV Data as Shown in Table 8.1 and Calculated PdV, ΣPdV, and Works of Hysteresis in Entrapment (W_T) and Extrusion-Reintrusion (W_{HY})

Pressure (PSI)	Intrusion (cc/gm)	Extrusion (cc/gm)	Reintrusion (cc/gm)	dV Intrusion	dV Extrusion	dV Reintrusion	Av. P (psi)	$P*dV$ Intrusion	$P*dV$ Extrusion	$P*dV$ Reintrusion
								(Joule/gm)	(Joule/gm)	(Joule/gm)
30	0.008	0.079	0.076	—	—	—	—			
50	0.024	0.082	0.077	0.016	0.003	0.001	40	0.004412608	0.000827364	0.000275788
100	0.0565	0.0853	0.0805	0.0325	0.0033	0.0035	65	0.014565054	0.001478913	0.001568544
200	0.06909	0.0865	0.084	0.01259	0.0012	0.0035	150	0.013020641	0.001241046	0.003619718
300	0.075	0.087	0.086	0.00591	0.0005	0.002	250	0.010186919	0.000861838	0.00344735
500	0.078	0.088	0.0878	0.003	0.001	0.0018	400	0.00827364	0.00275788	0.004964184
700	0.08	0.0882	0.0884	0.002	0.0002	0.0006	600	0.00827364	0.000827364	0.002482092
1000	0.081	0.0885	0.0885	0.001	0.0003	1E-04	850	0.005860495	0.001758148	0.000586049
2000	0.083	0.0895	0.0895	0.002	0.001	0.001	1500	0.0206841	0.01034205	0.01034205
4000	0.086	0.091	0.091	0.003	0.0015	0.0015	3000	0.0620523	0.03102615	0.03102615
6000	0.088	0.092	0.092	0.002	0.001	0.001	5000	0.068947	0.0344735	0.0344735
8000	0.0895	0.0928	0.0928	0.0015	0.0008	0.0008	7000	0.07239435	0.03861032	0.03861032
10000	0.091	0.0935	0.0935	0.0015	0.0007	0.0007	9000	0.09307845	0.04343661	0.04343661
20000	0.0953	0.0965	0.0965	0.0043	0.003	0.003	15000	0.44470815	0.3102615	0.3102615
30000	0.0985	0.0985	0.0985	0.0032	0.002	0.002	25000	0.551576	0.344735	0.344735

ΣPdV = 1.373620739 0.821810319 0.829553067

W_T = 0.55181042 (Intrusion–extrusion)

W_{HY} = 0.007742748 (Reintrusion–extrusion)

Units:

To convert PSI (lb_f/in^2) to dyne/cm^2 (gm-cm/sec^2),

Multiply by 6.8947E4;

$PV = (\text{gm cm/sec}^2 \cdot \text{cm}^2) * (\text{cm}^3/\text{gm})$

$= (\text{gm cm}^2/\text{sec}^2)(1/\text{gm})$

To convert (gm cm^2/sec^2) to Joules multiply by $10^{(-7)}$

178

can be described by one, two, or three axes, respectively. Fractal geometry relates to objects that fall between the linear and the planar and between the planar and volumetric and can be described by nonintegral, or fractal, dimension. In the fractal geometry, therefore, objects are defined by numbers between 1 and 2 when they are curved lines and between 2 and 3 when the objects are conventionally neither planar nor three-dimensional, such as pores in a rock.

Linear extensions, surfaces, and three-dimensional objects, when magnified or reduced in all directions and yet look the same, are called self-similar. For example, a 1 cm cube (volume = 1 cm^3) doubled in its spatial dimensions looks like the original cube even though its volume has increased to 8 cm^3 (2^3). The volume expansion factor, denoted by K, can be expressed in terms of length L, and the system dimension D as shown in Eq. (8.23). Self-similar objects obey what is called a power-law behavior and are said to be indicative of fractal geometry (Mandelbrot, 1982), expressed as

$$K = L^D \tag{8.23}$$

Taking the logarithm on both sides of Eq. (8.23) and rearranging, we get

$$D = \frac{\ln K}{\ln L} \tag{8.24}$$

The dimension D can take any value, integer or noninteger. When the values turn out to be noninteger, D is said to have a fractal dimension.

Reconsider the example of the cube given previously, but now that it has side length of R. Dividing this cube into p^3 smaller cubes and deleting some of the cubes, the remaining number of cubes can be designated as $N_{1(p)}$. If we repeat this operation n times, the sides of the cubes become

$$r_1 = \frac{R}{p}, \, r_2 = \frac{r_1}{p} = \frac{R}{p^2}, \cdots r_n = \frac{R}{p^n} \tag{8.25}$$

According to Eq. (8.24), L is given by

$$K = L^D = (p^n)^D = \left(\frac{R}{r_n}\right)^D = \left(\frac{r_n}{R}\right)^{-D} \tag{8.26}$$

The volume after n operations will be K times the original volume and can be written as

$$V_n \propto r_n^{3-D} \tag{8.27}$$

Pore distributions in a solid can be viewed as self-similar with the limiting case $n \to \infty$ or $r_n \to 0$, and their respective volume $V_n \to V(r)$. Following Friesen and Mikula (1987), they can be defined as

$$\frac{dV}{dr} \propto r^{2-D} \tag{8.28}$$

The pore volume is given by $V_{\text{pore}}(r) = R^3 - V(r)$; hence

$$-\frac{dV_{\text{pore}}}{dr} \propto r^{2-D} \tag{8.29}$$

where D is the fractal dimension, V is the intruded volume of mercury in pores of radius r, obtained from PV data by the application of the Washburn equation:

$$r = -\frac{2\gamma \cos\theta}{P} \tag{8.30}$$

or

$$r \propto \frac{1}{P}, \qquad dr = -\frac{dP}{P^2} \tag{8.31}$$

Using Eq. (8.31) with Eq. (8.29), we get

$$\frac{dV}{dP} P^2 \propto P^{D-2} \tag{8.32}$$

Taking the logarithm on both sides of Eq. (8.32) and using some algebraic manipulation give the following final expression:

$$\log\left(\frac{dV}{dP}\right) \propto (D-4) \log(P) \tag{8.33}$$

According to Eq. (8.33), when pore-size distributions plotted for $\log(dV/dP)$ versus $\log(P)$ give a straight-line plot, thus obeying a power law, the slope of the line is the

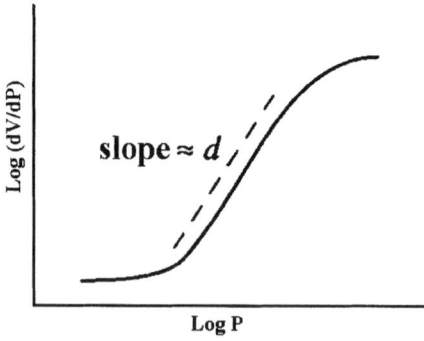

FIGURE 8.7   Fractal dimension. This plot of $\log(dV/dP)$ versus $\log P$, shows the *scaling range*, in which the pore-size distributions obey a power law. The slope of this line gives the fractal dimension d.

TABLE 8.3 The Mercury Intrusion Porosimetric (MIP) Test Data and Calculated Fractal Dimensions of Pore Volume (The plots (1–4 in right column) of the fractal dimensions at each stress range is shown in Fig. 8.8)

Mercury Pressure P (MPa)	Stress Range σ (MPa)				
	0–45	45–60	60–90	90–114	114–160
<1	2.357880	2.363411	2.366374	2.395242	2.471584 (1)
1–10	2.533292	2.546104	2.569615	2.582980	2.628543 (2)
10–100	2.652650	2.669237	2.671110	2.672543	2.689200 (3)
>100	2.711808	2.724178	2.731292	2.740267	2.762891 (4)
Average	2.557370	2.575735	2.584598	2.587785	2.637485

Source: Xie, H., Wang, J., and Qan, P., Fractal characters of micropore evolution in marbles, *Phys. Lett. A*, **218**: 275–280, 1996.

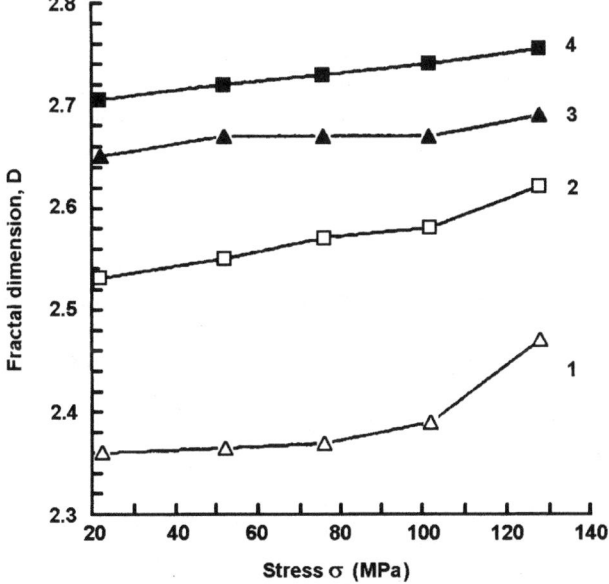

FIGURE 8.8 Fractally behaving segments of an irregular slope. The slope of each segment of the curves, e.g., curve 1, has its own fractal dimension *D*. *Source*: Reprinted from Xie, H., Wang, J., and Qan, P., Fractal characters of micropore evolution in marbles, *Phys. Lett. A*, **218**, 275–280, 1996. Reproduced with permission from Elsevier Science.

fractal dimension. However, pores in different size ranges in a porous rock often yield different slopes. In practice, the power law holds mostly for an intermediate (Fig. 8.7) range, or *scaling region*, of porosity. Sometimes, however, the power law may hold only for small individual segments of a pore-size distribution. Then, the average of various fractal dimensions is taken to describe all pore ranges (Table 8.3 and Fig. 8.8).

Let us describe how the data of Figure 8.8 and Table 8.3 were generated. Five marble samples were compressed in a testing machine at stress ranges shown in Table 8.3 to generate different pore sizes. Each sample was then injected with mercury in a porosimeter at four pressure ranges. The average fractal dimension of pore-size distribution of each sample is shown at the bottom of Table 8.3 while Figure 8.8 shows the irregularity of the respective slopes.

The idea behind presenting this example is that slopes of pore-size distributions are by no means smooth. Often one has to break down irregular slopes into segments, determine the fractal dimension of each segment of the slope, and then represent the entire distribution as an average slope. That is what we have done in Chapter 9 to show fractal dimensions of various Sphinx strata and use these to determine their durability factors.

SUMMARY

Mercury porosimetry consists of injecting mercury into the pores of stone under pressure. By this application, pressure–volume data are generated that can be expressed by (1) pore-size distributions, (2) a pore potential, and (3) fractal dimension. These properties, being unique to each rock type, can be used to characterize stone. We will apply these properties in Chapter 9 to determine durability factors for various limestone and dolostone.

REFERENCES

Anon., Understanding the bulk and skeletal density computations for the autopore and pore sizer, *Micromeritics*, Norcross, Georgia, Application Note Number 20, (n.d.).

Anon., Pore surface area computation from mercury penetration data by work of immersion and the assumption of cylindrical geometry, *Micromeritics*, Norcross, Georgia, Application Note Number 31 (n.d.).

Friesen, W. I. and Mikula, R. J., Fractal dimension of coal particles, *J. Colloid Interfacial Sci.*, **120:** 263–271, 1987.

Lowell, S., and Shields, J. E., *Powder Surface Area and Porosity*, Chapman and Hall, New York: 1981, p. 234.

Mandlebrot, B. B., *The Fractal Geometry of Nature*, Freeman: San Francisco, 1982, p. 460.

CHAPTER 9

Durability of the Sphinx Limestone

9.1 INTRODUCTION

The Sphinx is a monolith (Fig. 9.1) created nearly 5,000 years ago when the limestone strata around it had been quarried for construction of Kephren's Pyramid and other pharaonic monuments. The original vertical profile of the Sphinx must have been a smooth curve. The modern profile, however, consists of alternating projections and recessions (Fig. 9.2), revealing a differential weathering of the strata over time. The origin, properties, and structure of these strata are discussed, as they affect the durability of the Sphinx.

The durability of limestone is often considered as its ability to withstand pressure generated in the pores. Of the several factors that control durability discussed in Chapter 4, pore-size distribution was considered to be preponderant. Assigning empirical durability factors to some of the Sphinx strata, the relative weathering of which could be judged by observation, model equations are developed in this chapter for prediction of durability factors based upon regression, fractal analysis, and the pore potential of pore-size distributions, discussed in Chapter 8.

9.2 ORIGIN, STRUCTURE, AND SEDIMENTARY FEATURES OF THE SPHINX STRATA

9.2.1 Origin

The Great Sphinx—there are many other small sphinx structures in Egypt—is located on the southeastern slope of the Giza Plateau on the west bank of the River Nile at Cairo. The limestone forming the Giza Plateau, as well as those at the hills on the eastern bank of Nile, belong to the Mokattam formation of the middle Eocene age. The strata of the Mokattam formation were laid down as horizontal layers upon the floor of a shallow sea.

The basin in which the Sphinx limestone was deposited can be visualized as a calm-water lagoon enclosed between a landmass on the west and a reef in the east growing on the continental shelf. Somewhat protected from waves by the reef, the lagoon was covered with mats of algae and other microscopic organisms containing fine needles of calcium carbonate as their skeletons. Upon the death of these

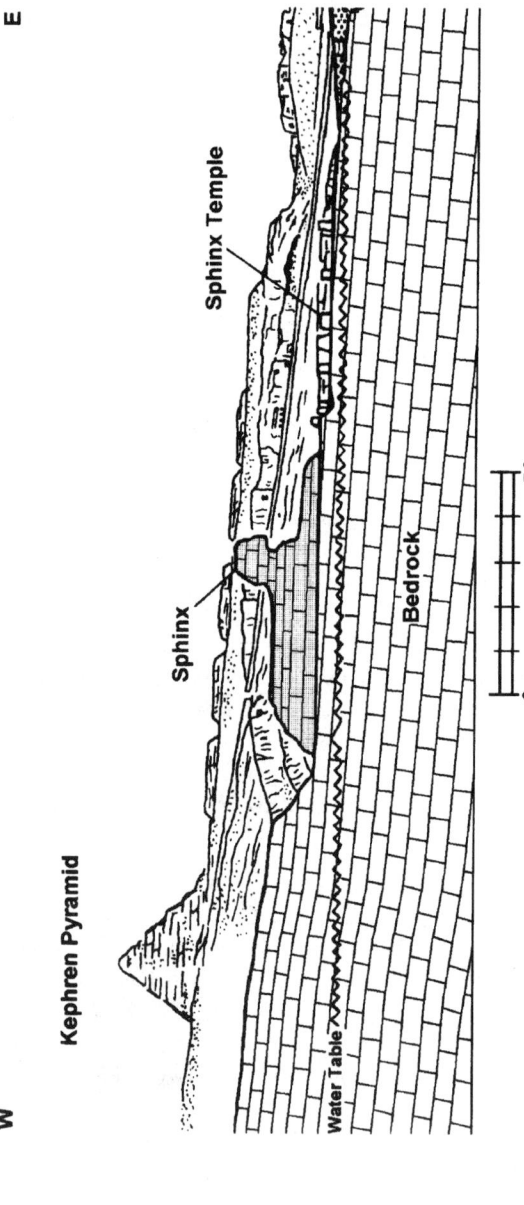

FIGURE 9.1 The Sphinx, a monolith. Carved in the Mokattam formation of Eocene age, the Sphinx is a structure remaining after the limestone beds around it were quarried. Some of the limestone was used in the construction of Kephren's Pyramid and several nearby temples. *Source:* Gauri, K. L., Sinai, J. J., and Bandyopadhyay, J. K., Geologic weathering and its implication on the age of the Sphinx, *Geoarchaeology*, **10**(2): 119–133, 1995. Reprinted by permission of John Wiley & Sons, Inc. Copyright 1995 John Wiley & Sons, Inc.

FIGURE 9.2 Weathering profile of the Sphinx. The strata of the Sphinx thorax are differentially weathered due to their variable makeup. The head, however, was carved in a highly durable rock of rather uniform composition. The paws of the Sphinx had weathered so much even during the pharaonic times, soon after the Sphinx had been carved, that construction of the limestone casing wall was found necessary to protect the bedrock.

organisms, the soft tissue decayed and the aragonite needles accumulated on the sea floor as carbonate mud, called biomicrite, which later was consolidated to form the Mokattam formation.

The seawater, particularly that outside the lagoon, was teaming with animal life consisting mainly of free-floating organisms, such as *nummulites* and many bottom-dwelling communities of *fusulinids*, *echinoderms*, and *molluscs*. The shells of these organisms were also made of carbonate minerals. As these organisms died, their shells settled upon the sea floor. Worked by waves, the shells were broken down into coarse-grained sand made of what are called allochemical particles. At times of heavy storms some of these particles were washed into the fine carbonate mud being deposited in the sheltered lagoon.

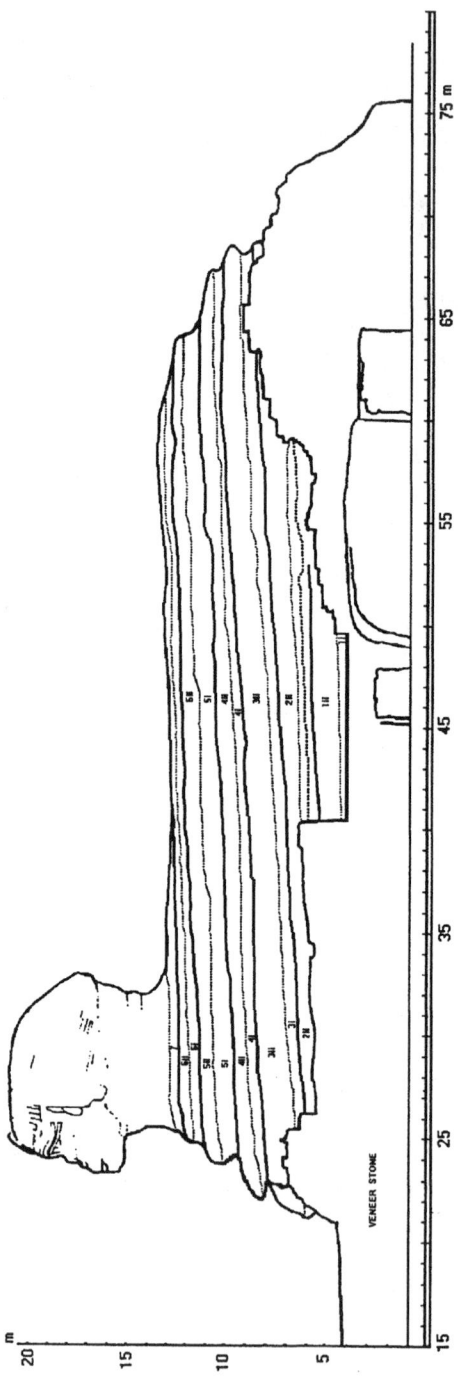

FIGURE 9.3 Longitudinal profile of the Sphinx. The strata of the thorax are marked 1 through 7 to identify the samples used in various studies reported in this chapter. *Source: Gauri, K. L. et al.*, Geologic features and durability of limestones at the Sphinx, reprinted from, *Engineering Geology of Ancient Works, Monuments, and Historic Sites*, P. G. Marinos and G. C. Koukis (eds.), Rotterdam: Balkema, 1988, pp. 723–729.

The sediments of the Mokattam formation were thus laid down as alternating layers of fine mud with sparse allochemical particles and mud containing large quantities of allochem. These sediments, when consolidated into rocks are termed, respectively, sparse biomicrite and packed biomicrite (Chap. 1).

As the carbonate minerals were being laid down, some noncarbonate matter was being washed into the lagoon from the adjacent coastal area (Chap. 2). This matter consisted of water-soluble minerals such as halite and gypsum and nonsoluble particles of quartz and clay minerals. These minerals are partly responsible for the deep weathering of some of the Sphinx strata.

The Mokattam formation was deposited on a gently sloping surface that dipped toward the southeast. Such a slope is called a paleoslope and can be reconstructed by means of stratum contours, which are lines joining points of the same elevation marked at the intersection of the upper surface of a formation with the land surface of today. The paleoslope and the strata of the Mokattam formation dip at an angle of nearly 4° to the southeast (Figs. 9.3 through 9.5).

After the Mokattam formation had been deposited, the entire region experienced tectonism in the form of epeirogenic movement, defined as broad uplift of a land area or sea bottom in which the strata remain parallel. Then, the subsurface water of the continent began circulating through the pores, dissolving some of the carbonate minerals, and creating a network of ink-bottle pores in the rock. In addition, some large caverns were created along joints (Chap. 11) where movement of large quantities of freshwater was possible.

We will now describe in detail the structural and sedimentologic features of the rock units of the Mokattam formation exposed at the Sphinx and their influence upon durability.

9.2.2 Structural Features

The structural features of the Mokattam formation date back to a tectonic event during the Cretaceous age that occurred before this formation was deposited. The features created by that episode were minor folds and joints.

The modern regional attitude of the bedding along the Nile Valley is a gentle northward dip so that the strata are progressively younger from Aswan towards Cairo. At Cairo, however, the strata dip toward the south (Fig. 9.6), indicating the presence of a local anticline that resulted from the Cretaceous tectonism that bent the strata into a series of anticlines with a northeastern trend and also created joints and faults that follow a NW–SE direction as well as parallel to the fold axis. The southern limb of that anticline is the paleoslope of the Mokattam formation in this area and the joints in it are the vertical joints that were able to emerge from below during eperiogenesis.

The density contour of poles to joints (see Chap. 3) measured in the rocks at the Sphinx and those forming walls of the Sphinx Ditch are shown in Fig. 9.7. Since the joints extend through the three members of the Sphinx strata (Fig. 9.4), their attitude appears to be consistent, and the data therefore have been combined in a single diagram. Figure 9.7 shows further that the joints fall into two orthogonal groups. One

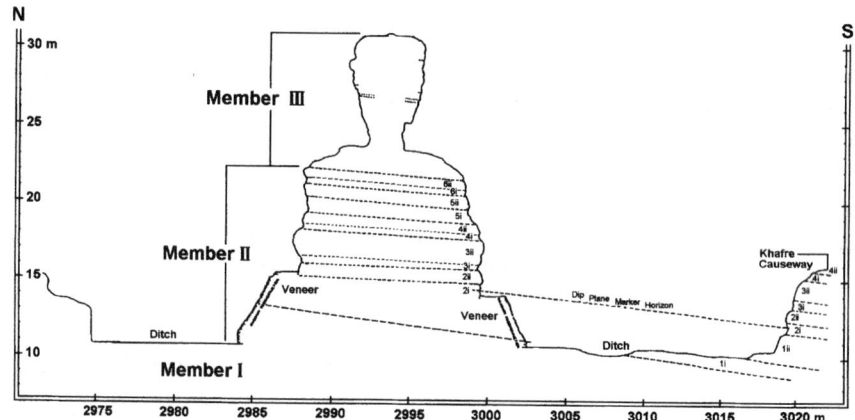

FIGURE 9.4 Dipping Sphinx strata. Beside other dipping beds, this profile of the Sphinx and the wall on the south shows bed 2ᵢ, which was used to reconstruct the paleoslope (Fig. 9.5) on which the Mokattam formation was deposited. *Source*: Gauri, K. L., Geologic study of the Sphinx, American Research Center in Egypt, *Newsletter*, **127**: 24–43, 1984. Courtesy ARCE.

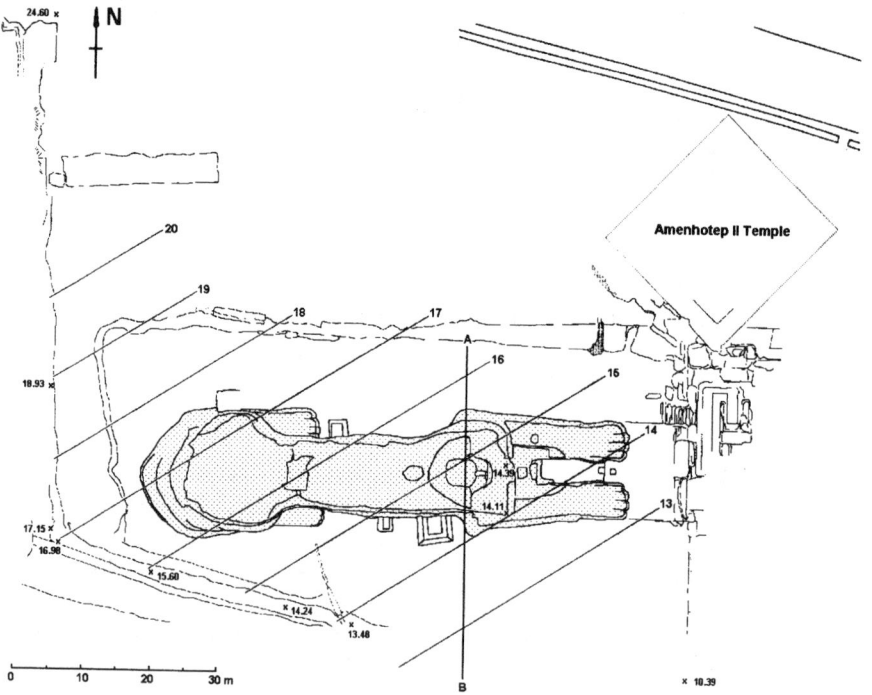

FIGURE 9.5 Stratum contours (meters above sealevel). Constructed for the upper surface of bed 2ᵢ, showing a dip of the paleoslope of nearly 4° southeast. *Source*: Gauri, K. L., Geologic study of the Sphinx, American Research Center in Egypt, *Newsletter*, **127**: 24–43, 1984. Courtesy ARCE.

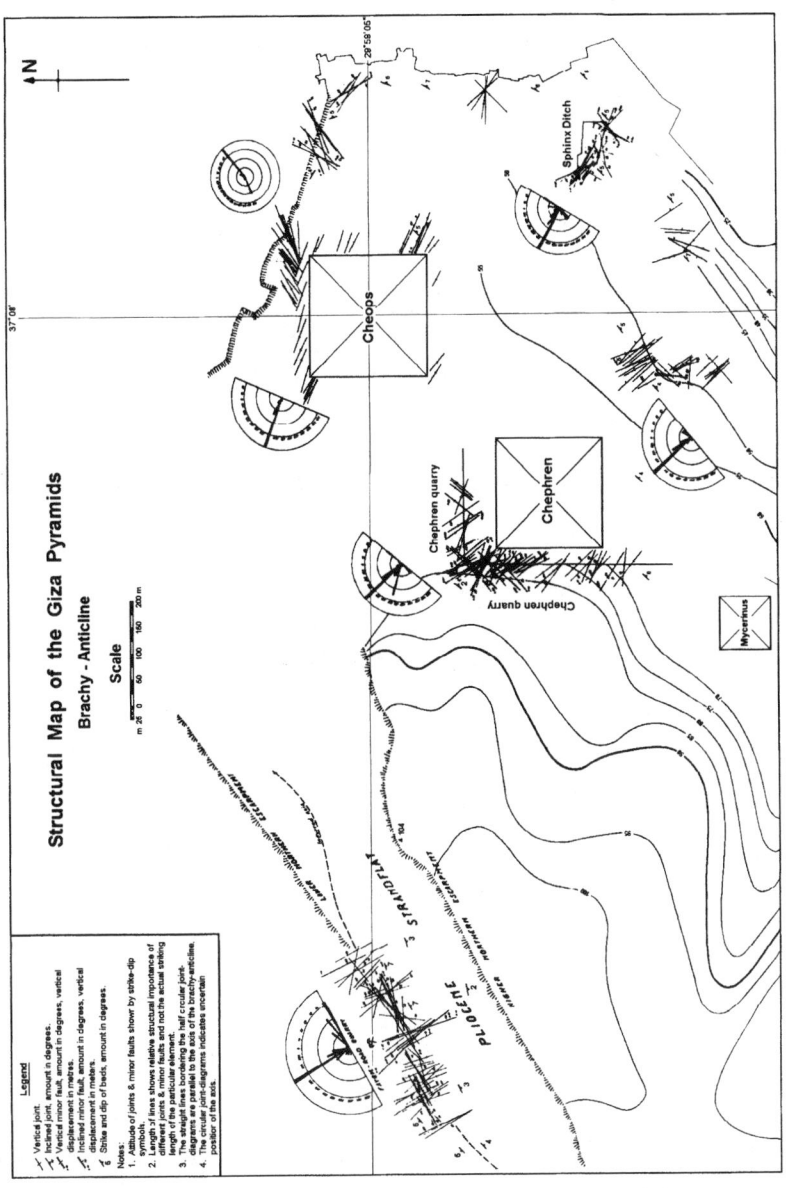

FIGURE 9.6 Structural map of the Giza Pyramids showing anticlines and joints. *Source:* Omara, S. M., The structural features of the Giza Pyramids area, thesis, Fouad L. University, Cairo, 1952. p. 84.

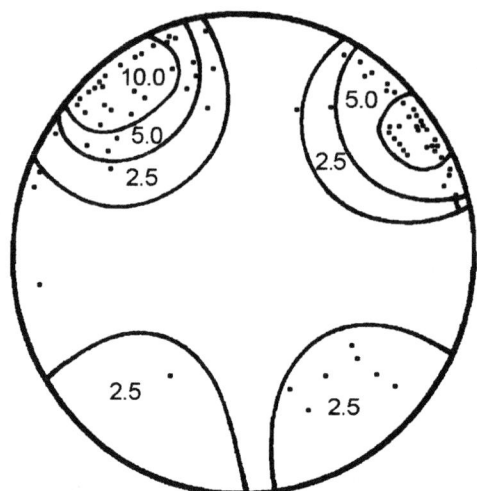

FIGURE 9.7 Joints in the Sphinx strata showing poles to 84 joints and density contours. Notice two orthogonal systems of nearly vertical joints trending northwest and southeast. *Source*: Gauri, K. L., Geologic study of the Sphinx, American Research Center in Egypt, *Newsletter*, **127**: 24–43, 1984. Courtesy ARCE.

of these groups trends generally northeast to southwest while the other trends northwest to southeast. The centers of these clusters are precisely perpendicular, with trends N36W and N54E. This pattern is also seen in the map showing the larger region of the Giza Plateau (Fig. 9.6).

The existence of the joints affects the durability of the Sphinx strata. We will describe this effect in the hard-limestone units where the joints are well developed, hence can be easily seen, and in the soft-limestone units where the joints feather out into a multitude of closely spaced fractures and are not clearly delineated.

In the hard-limestone units, the joints are often widely spaced (Fig. 9.8) so that the rock between joints weathers by the gradual separation of grains. However, when the joints are somewhat closer in such units, a spheroidal pattern of weathering appears (Fig. 9.9), the portion of strata closer to joints being weathered more intensively than those farther away. Where intersection of two NW- and SE-trending joints occur in such rocks or where the joints run subparallel with the outcrop, wedge-shaped blocks of rock are likely to separate after extensive solution exposure along joints. In Figures 9.2 and 9.8 such wedges can be seen ready to dislodge.

In the softer units, however, a multitude of microfractures is present although they are not clearly visible because of their infinitely small size. Weathering along these microfractures is highly intense and the rock pulverizes into fine dust. Figure 9.9 shows the softer limestone layer beneath the layer that shows spheroidal weathering. The large mass of fine debris visible at the base of the slope is largely the weathered product of the soft layer having accumulated in a short period of less than ten years as we have observed.

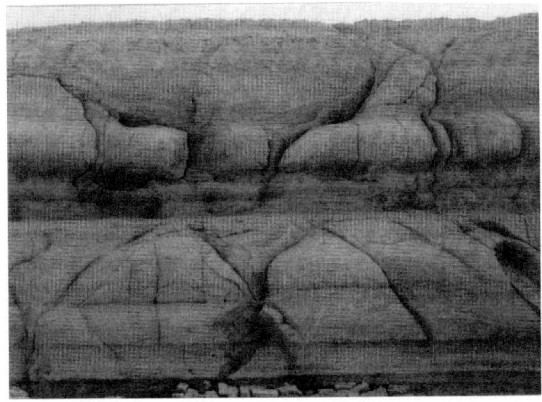

FIGURE 9.8 Joints in the Sphinx strata showing widely spaced joints in competent limestone strata and their apparent absence in interlayered marly strata. Notice the rock fragments about to fall off where joints intersect the outcrop surface.

FIGURE 9.9 Spheroidal pattern of weathering developed in the south wall of the Sphinx in a hard limestone unit in which the intersecting joints are closely spaced. Also pulverized soft rock beneath can be seen in which the joints feather out into closely spaced microfractures. *Source*: Gauri, K. L., Sinai, J. J., and Bandyopadhyay, J. K., Geologic weathering and its implication on the age of the Sphinx, *Geoarchaeology*, **10**(2): 119–133, 1995.

9.2.3 Sedimentary Features

The strata exposed at the Sphinx can be divided into three members based upon their sedimentary makeup:

- The lower member, forming the base of the Sphinx, is a bioherm made of a massive bed of consolidated skeletal debris. This member is largely covered by stone blocks (Fig. 9.2, lower right) used in many restorations of the Sphinx. Although quite durable, these strata at the Sphinx were covered by veneer stone to protect the lower layer of the overlying members that had highly weathered.

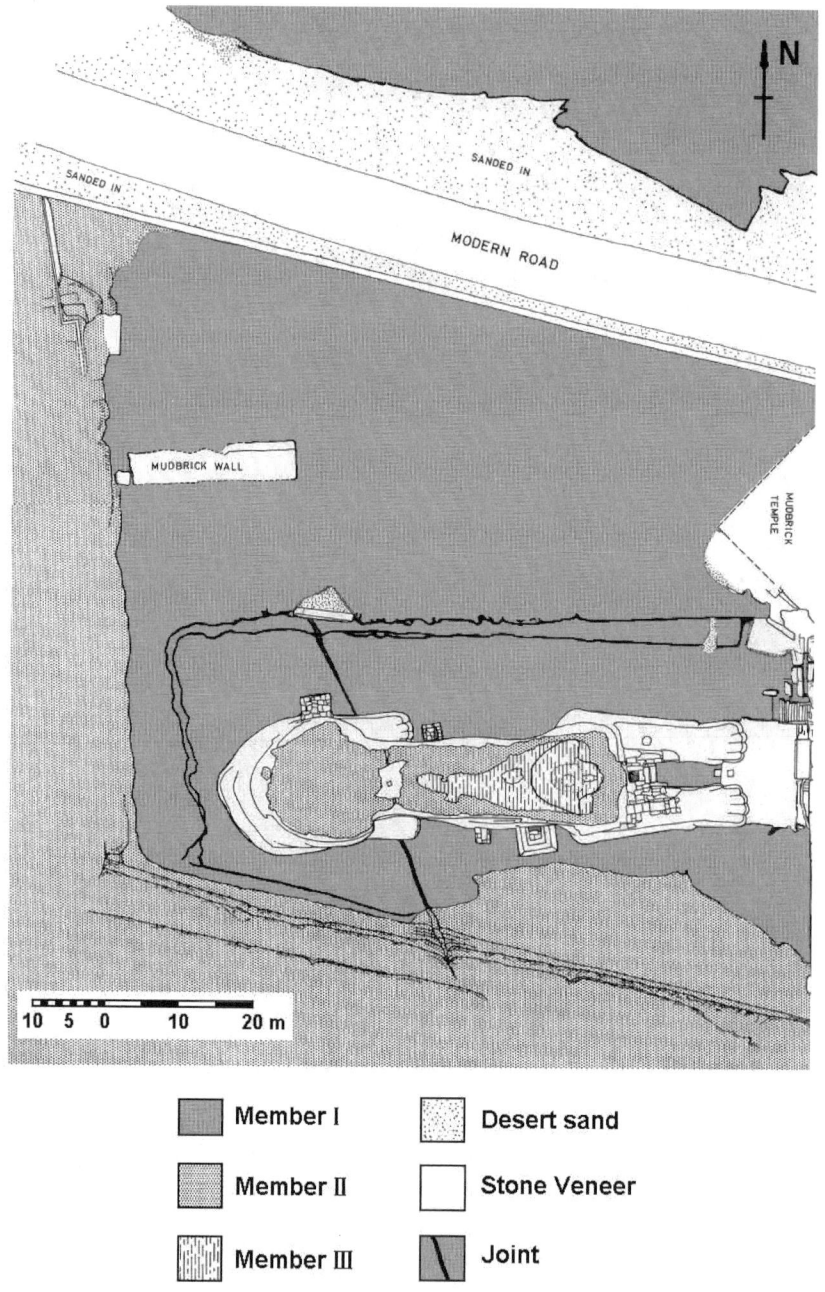

FIGURE 9.10 Geologic map of the Sphinx showing distributions of the rock members at the Sphinx and vicinity. Notice member I buried underneath member II in the ditch south of the Sphinx due to the southeasterly dip of the strata. Notice also a NW–SE joint system passing through the rear portion of the Sphinx. *Source*: Gauri, K. L., Geologic study of the Sphinx, American Research Center in Egypt, *Newsletter*, **127**: 24–43, 1984. Courtesy ARCE.

- The middle member consists of strata that form the thoracic region of the Sphinx. These strata are composed of seven alternating beds of micritic limestone. The sequence begins at the bottom with a marly (clastic plus micrite) layer with sparse shells grading upwards into allochemical micrite. In fact, the strata of the middle member show a sequence of rhythmically deposited beds, the lower portion of each bed being an impure, softer limestone while the upper portion is cleaner and harder limestone (Fig. 9.8). In general, however, the strata up to the fourth from the bottom can be designated as sparse biomicrite while the remaining upper four layers as packed biomicrite (Chap. 1, Figs. 1.6 and 1.7).

- The geologic profile shows the relative thickness of the beds and the differential weathering of the middle member. The units of the thorax are designated as 1_i, 1_{ii} to 7 (Figs. 9.3 and 9.4). Layers with the subscript designation of i are the lower, more weathered marly beds and those labeled ii are relatively purer, less weathered limestone. The purity of the limestone increases in the ascending direction so that bed 7 is undifferentiated clean limestone, which shows the least amount of weathering as compared with all other rocks of the thorax.

- The strata present in the neck and the head region consist of a lower, highly micritic sequence and upper allochemical-rich layers, similar to the layers of the thoracic region. But these strata are much thicker than those of the middle member. The massive strata of the head, however, have thin intercalation of softer limestone (Fig. 9.2).

These strata are shown in the geological map of the Sphinx area (Fig. 9.10).

Geologic maps in general show the regional distribution of exposed strata as projected upon a horizontal plane. Thus the strata of the upper member of the Sphinx (neck and head), which are not present elsewhere in the region, appear only in the front-central part of the Sphinx, followed outward by the strata of the middle member (thorax). The lower member is present mainly north of the Sphinx and adjacent to it in the south, being buried farther south and west under the middle member due to the southeasterly dip of the strata.

9.3 NONCARBONATE MINERALS IN THE SPHINX LIMESTONE

Noncarbonate minerals in carbonate rocks affect their durability. Noncarbonate minerals are commonly distinguished as chemical precipitates and clastic particles. The chemical precipitates, also called evaporites, are originally the water-soluble salts carried as ions or as solid particles in water into the carbonate basin from adjacent land or supratidal sites. Clastic particles are particles of minerals, such as clay and sand, which are transported from land by wind or water into the calm water of carbonate basin. We will discuss in the following the composition of these foreign materials and their effect upon the durability of the limestone.

9.3.1 Water-Soluble Salts

Water-soluble salts are commonly present in small quantities in the limestone and therefore are difficult to identify as part of the bulk rock. However, these salts can be dissolved from the limestone by placing a sample in deionized water and then slowly evaporating the water so that the salts crystallize. Two major salts from the Sphinx samples have been identified by X-ray diffraction as gypsum ($CaSO_4 \cdot 2H_2O$) and halite (NaCl). Figure 9.11 shows the crystals of these salts growing from the solution as revealed by light microscopy. The ionic composition of solutions extracted from samples of the various layers of the Sphinx is shown in Table 9.1. This table indicates that more weathered strata have generally a larger content of water soluble salts. In evaluating the correlation of salt distribution with the degree of weathering certain points must be kept in mind:

- A large quantity of gypsum ($CaSO_4 \cdot 2H_2O$)—indicated by the presence of sulfate ions (SO_4^{2-}) in Table 1—promotes stone disintegration, but when present in a small quantity, gypsum is able to form a thin crust in desert climate, called *duricrust*, which may in fact protect the underlying stone from future decay. For

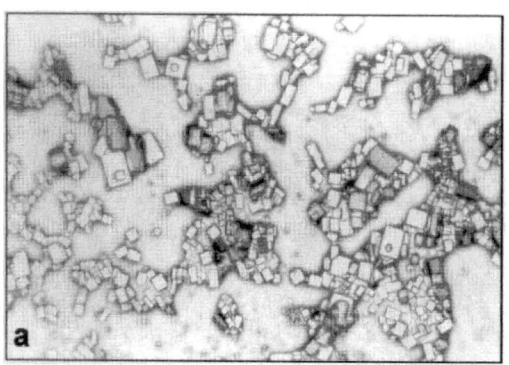

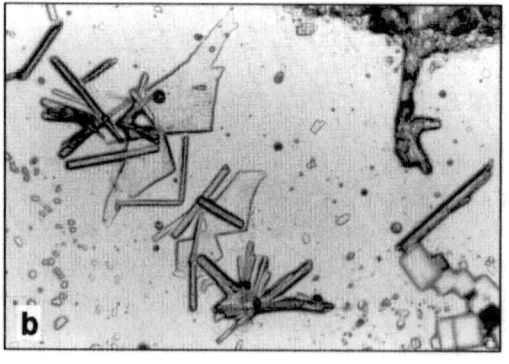

FIGURE 9.11 Water-soluble salts in the Sphinx strata. These salts were extracted in solution by placing a rock sample in deionized water. Drops of the solution were then slowly evaporated upon a glass slide and photographs of the growing crystals made under light microscopy. (a) shows cubic halite crystals and (b) shows bladed gypsum crystals. *Source*: Gauri, K. L., Geologic study of the Sphinx, American Research Center in Egypt, *Newsletter*, **127**: 24–43, 1984. Courtesy of ARCE.

TABLE 9.1 Ionic Composition (Wt. Percent of Sample) of Salts Extracted in Deionized Water from Sphinx Beds

Bed	K^+	Na^+	Ca^{2+}	Mg^{2+}	NO_3^-	Cl^-	SO_4^{2-}
6_{ii}	0.009	0.074	0.061	0.004	0.014	0.098	0.998
6_I	0.012	0.163	0.269	0.021	0.068	0.270	0.726
5_{ii}	0.012	0.096	0.269	0.021	0.057	0.220	0.726
5_I	0.034	1.110	0.267	0.038	0.166	2.050	1.110
4_{ii}	0.009	0.130	0.110	0.030	0.049	0.370	0.038
4_i	0.009	0.210	1.480	0.021	0.013	0.410	2.900
3_{ii}	0.018	0.600	0.520	0.017	0.014	1.130	1.090
3_i	0.012	0.380	0.085	0.015	0.019	0.700	0.058
2_{ii}	0.006	0.840	0.091	0.008	0.006	1.560	0.130
2_i	0.028	0.300	0.480	0.042	0.064	0.550	1.170
1_{ii}	0.008	0.500	0.071	0.014	0.008	1.000	0.062
1_i	0.020	1.130	0.064	0.020	0.017	2.180	0.086

Source: Gauri, K. L., Geologic study of the Sphinx, American Research Center in Egypt, *Newsletter*, **127**: 24–43, 1984. Reprinted by courtesy of ARCE.

example, the face of the Sphinx is well preserved because it has a protective crust due, presumably, to a minor quantity of gypsum in the parent stone. Further, limestone blocks used in the restoration of the Sphinx in pharaonic times seem to have not weathered at all because of the protective crust and suitable pore-size distributions that we will discuss later.

- Halite (NaCl) is highly disruptive in any quantity because, among other reasons discussed later, it is unable to form a protective crust even in a desert climate due to its high solubility in water.

The water-soluble salts are disruptive only when they are concentrated in a large quantity near the surface of the stone. Originally, they are rather sparsely distributed throughout the rock mass; however, they are relatively highly concentrated towards the surface in the strata of the Sphinx. This concentration near the surface can be explained as follows.

The Sphinx had been buried under the desert sand for long periods of time during its existence of nearly 5,000 years. During the periods of burial, the rock became wet to a considerable depth as can be seen today wherever the sand is piled up against the bedrock (Fig. 9.12). This wetting is due to the atmospheric moisture that condenses upon the sand, fills the pores between the sand grains, and is then pulled in the rock by capillary action. Later, when the sand burying the Sphinx was removed during several restorations, the salts became accumulated near the surface as the bedrock dried.

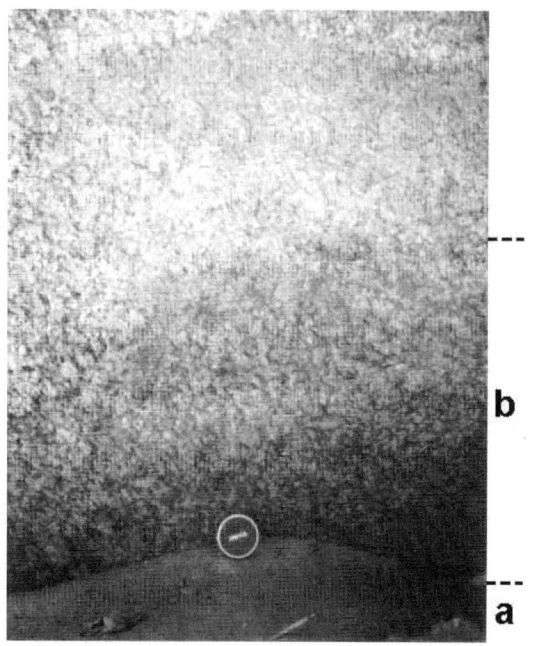

FIGURE 9.12 Wetting of rock in the desert sand. The sand condenses moisture in the air when it becomes cool in the night. The salts in the sand aid in the condensation due to their deliquescence. The condensed moisture is then wicked into the rock by capillarity. In this photograph label a marks the surface of sand upon which a solid rock is resting and is saturated with water in the horizon b.

FIGURE 9.13 Fog in the desert showing Kephren's Pyramid enveloped in the fog on an early December morning. The rock surfaces in the area at that time were drenched with the condensed moisture.

Salts are generally the major cause of masonry destruction in the entire world, but they are most effective in arid climates. Here, even the sparse amount of available moisture is able to condense in the atmosphere as thick fog in the early hours of the morning when climatic conditions are suitable. Figure 9.13 shows such a fog on an early December morning. The fog settles upon the stone and condenses as droplets of water on the cooled stone surface. The drops of water form a concentrated salt solution that then enters the stone and dissolves more salt already present in the pores. At sunrise, as the water begins to evaporate and the crystals of salt grow, crystallization pressure is produced. This phenomenon of moisture condensation (Appendix B) occurs quite frequently in the desert climate, even though dense fogs may be infrequent. In the morning one can often hear the sound of popping stone resulting from the pressure generated under the surface layers.

The arid climate is further conducive to stone disintegration by salt crystallization because the desert sand is, in general, enriched in salts, which in humid climates are drained away into the world oceans. Some of these salts, halite, for example in the case of the Sphinx, are deliquescent; they convert moisture to liquid water at a relative humidity less than that needed for the dew point.

9.3.2 Clastic Minerals

Quartz particles and clay minerals are the major clastic particles present in the Sphinx limestone. Like water-soluble salts, they are also sparsely present, but can be isolated first by leaching the water-soluble minerals in deionized water and then by the

TABLE 9.2 Proportion of Carbonate to Clastic Minerals in Strata of the Sphinx

		Clastic Particles (%)			
Bed	Carbonate Fraction (%)	Sand 60–2,000 μm	Silt 4–60 μm	Clay <4 μm	Total
6_{ii}	95.654	2.457	1.158	0.611	4.225
6_{I}	95.018	7.324	0.625	1.515	9.464
5_{ii}	95.877	1.957	0.899	1.140	3.991
5_{I}	88.072	4.605	2.775	4.067	11.447
4_{ii}	94.342	0.030	4.530	0.370	4.920
4_{I}	86.817	0.160	7.330	0.650	8.140
3_{ii}	93.981	0.400	1.700	0.510	2.630
3_{I}	93.501	0.050	4.720	0.460	5.230
2_{ii}	93.710	0.050	2.890	0.710	3.650
2_{I}	88.060	0.140	8.480	0.690	9.310
1_{ii}	91.757	0.240	5.380	0.950	6.580
1_{I}	70.353	0.200	25.170	0.760	26.130

Source: Gauri, K. L., Geologic study of the Sphinx, American Research Center in Egypt, *Newsletter*, **127**: 24–43, 1984. Courtesy of ARCE.

TABLE 9.3 X-Ray Diffraction Analysis of Clay Minerals Concentrated from Limestone Samples

2θ (deg)	d Spacing	Mineral
5.1–5.3	17.30–16.66	Montmorillonite
7.2–7.5	12.44–8.78	
9.8–10.2	9.02–8.66	Illite
12.5	7.08	Kaolinite
25.0–25.2	3.56–3.53	

digesting the carbonate fraction in dilute hydrochloric acid as discussed in detail in Chapter 3. The clastic minerals can then be separated into size fractions by sieving. Table 9.2 gives the proportions of the carbonate fraction and the clastic minerals in various strata of the Sphinx.

X-ray diffraction of sand and silt fraction revealed the presence of quartz only. The clay minerals are identified in Table 9.3. Kaolinite and illite were present in all samples, with kaolinite being the preponderant species. Montmorillonite appeared in a small quantity in bed 3 only. The identification of these minerals on a dry oriented sample, as shown in Table 9.3, was confirmed by other treatments, such as glycolocation given in Chapter 3.

Among these clay minerals montmorillonite is the only type that would expand upon hydration and thus be the cause of stone disintegration. Since it is absent from most strata and when present it is sparse, its effect upon durability of Sphinx can be expected to be minimal. However, this and the other clay minerals affect durability indirectly by altering the pore-size distributions and should be considered in the overall weathering of limestone.

9.4 PORE-SIZE DISTRIBUTIONS AND DURABILITY OF THE SPHINX LIMESTONE

Pore-size distributions determine the magnitude of pressure developed by salt crystallization because stresses due to crystallization are mainly based upon the chemical potential of a crystals growing in pores (Chap. 4). In interconnected small and large pores, the change in pressure can be given by

$$dp = 4\gamma\left(\frac{1}{d} - \frac{1}{D}\right) \tag{9.1}$$

where γ is the surface tension between solid and liquid (90 dyn/cm for sodium chloride solution) and d and D are the diameters of small and large pores. Table 9.4 gives the

TABLE 9.4 Pore-Size Distributions and Fractional Pore-Volume (Percent of Total Pore Volume) in the Sphinx Strata

Beds	V_1 (>5 µm)	V_2 (0.5–5 µm)	V_3 (<0.5 µm)
7	74.68	9.09	16.23
6_{ii}	74.59	8.84	16.57
6_I	69.3	11.25	19.54
5_{ii}	55.0	19.12	25.88
5_I	52.11	18.77	29.12
4_{ii}	74.88	5.03	20.09
4_I	71.49	4.34	24.17
3_{ii}	57.56	13.04	29.4
3_I	57.23	3.7	39.07
2_{ii}	66.01	8.04	25.95
2_I	43.86	3.32	52.82
1_{ii}	63.28	4.0	32.71
1_I	40.69	5.5	53.81
PH	49.68	24.65	25.66

Source: Gauri, K. L., and et al., Geologic features and durability of limestones at the Sphinx, reprinted from *Engineering Geology of Ancient Works, Monuments, and Historic Sites*, P. G. Marinos and G. C. Koukis (eds.), 723–729. Rotterdam: Balkema, 1988, pp. 723–729.

Note: Beds 1–7 are shown in Figure 9.3; PH is a sample from a highly durable limestone block used in the pharaonic restoration of the Sphinx.

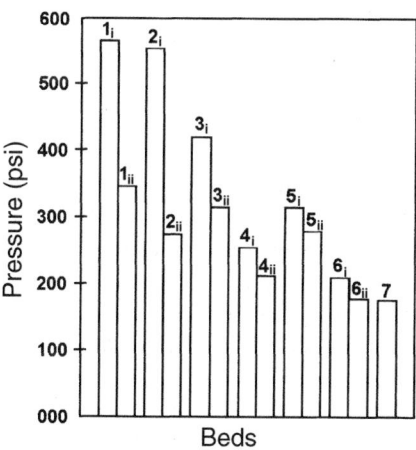

FIGURE 9.14 Crystallization pressure in the Sphinx strata showing the greater pressure developed in rocks with larger microporosity (Table 9.4). Notice that the pattern of pressures follows the rhythmic development of the strata. The crystallization pressures were calculated by the method shown in Chapter 4. *Source*: Gauri, K. L., et al., Geologic features and durability of limestones at the Sphinx, reprinted from, *Engineering Geology of Ancient Works, Monuments, and Historic Sites*, Marinos, P. G., and Koukis, G. C. (eds.), Rotterdam: Balkema, 1988, pp. 723–729.

pore-size distributions in various strata of the Sphinx and the calculated crystallization pressures are shown in Figure 9.14 (see Chapter 4 for sample calculation).

Pore-size distributions show more clearly a good correlation with durability than other properties, such as salt and clay-mineral content. The rocks with a larger volume of small pores are clearly more susceptible to weathering than those that have a greater abundance of large pores (Fig. 9.15). On the basis of pore-size distributions we can now develop durability factors for the Sphinx strata.

9.4.1 Durability Factors

Durability factors (DF) are numbers in an empirical scale by which the durability of a rock can be determined as compared with a standard. In the case of the Sphinx, for example, the well-preserved stone used for restoration during pharaonic times (PH, Table 9.4) and almost equally well-preserved stratum 7 can be assigned DF 10 and 9.5, respectively, and the most highly weathered stratum 1_i assigned DF 1. Other strata of the Sphinx can now be given DF values and checked against the relative weathering they have experienced.

In concept, the DF are similar to the Mohs scale of hardness of minerals (Chap. 1) in which the hardness index (1 to 10) is assigned upon the ability of a mineral to scratch the other, even though the underlying cause such as the atomic properties and crystal structure are unknown. Furthermore, in the Mohs scale, as in the DF, it is not known in which scale (linear, fractional, exponential) one group is related to the other.

However, unlike Mohs' scratch test that must be performed each time when the hardness of a mineral needs to be determined, the durability factors cannot be determined for all limestone types because a direct comparison is not possible. Therefore, DF of the standard rock types must be correlated statistically with some intrinsic property of the standard. In Chapter 4 we indicated that pore-size distributions have the preponderant effect upon durability. Now, we will use the pore-size distributions of the selected set of rocks to obtain model equations that can predict durability of other limestone when their pore-size distributions are known.

Pores can be related to durability in a variety of ways. The simplest case is for pore-size distributions of the standards correlated with their DF by regression techniques. Another method for correlation is the resistance of pores to intrusion of mercury under pressure, expressed as the pore potential. Yet another method can be the pore structure itself, considered as a fractal dimension. In all these methods, however, independent relations of the pore-size distribution and the empirical DF must be developed and confirmed by application to other known strata, the relative durability of which can be compared with the standards in the field.

9.4.1.1 *Durability Factors by Regression Analysis* The statistical relation of pore-size distributions with DF can be expressed by an equation of the form:

$$A_1V_1 + A_2V_2 + A_3V_3 = DF \tag{9.2}$$

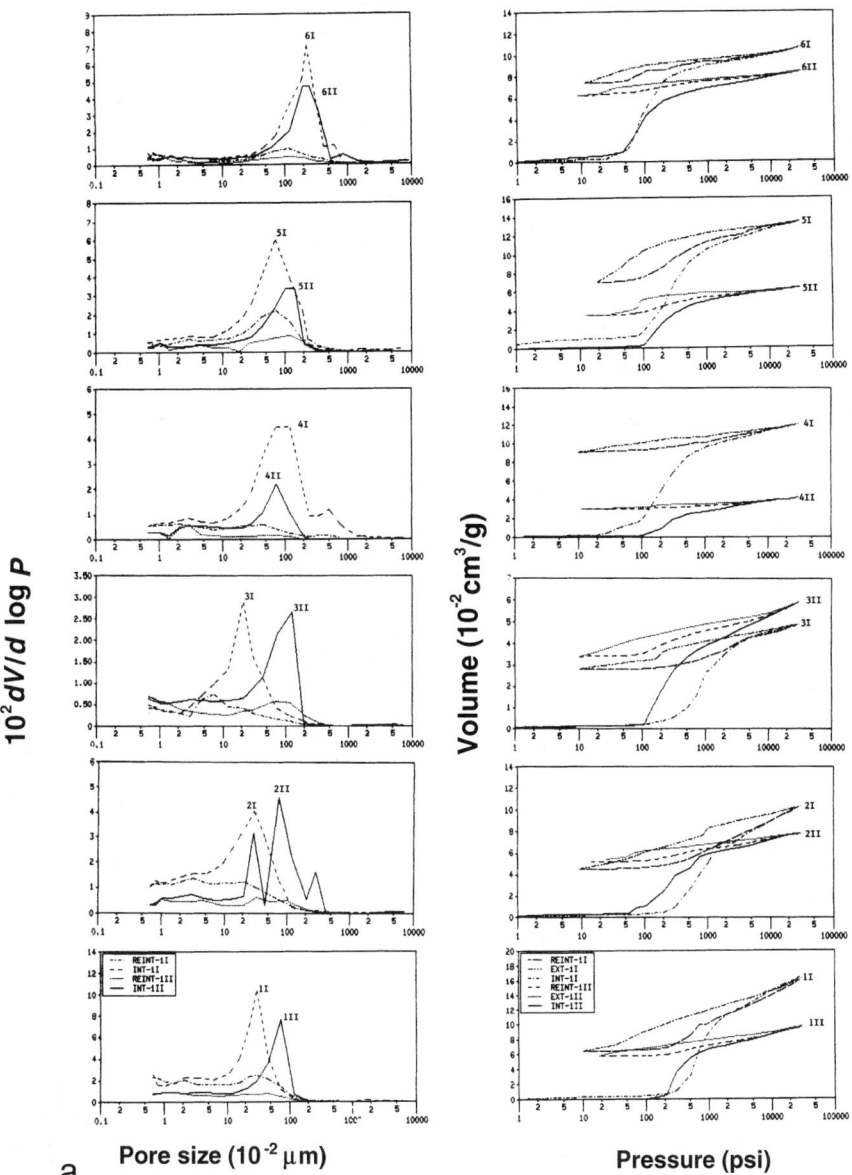

a   **Pore size (10⁻² μm)**

Pressure (psi)

FIGURE 9.15 Relationship of small and large pores in the Sphinx strata (a) shows the incremental initial intrusion (INT) and the reintrusion (REINT) curves generated in mercury intrusion porosimetry. The reintrusion curves show the volume of small pores while the initial intrusion curve shows the combined volume of small and large pores. (b) Cumulative curves show that the volume of small pores decreases with the ascending order of the strata, which correlates well with the durability factors. *Source*: Bandyopadhyay, J. K., Yerrapragada, S. S., and Gauri, K. L., Artificial neural networks and the durability of Sphinx limestone, *J. Mater. Civil Eng.*, **7**(3): 174–177, 1995. Reprinted with permission, copyright 1995 ASCE.

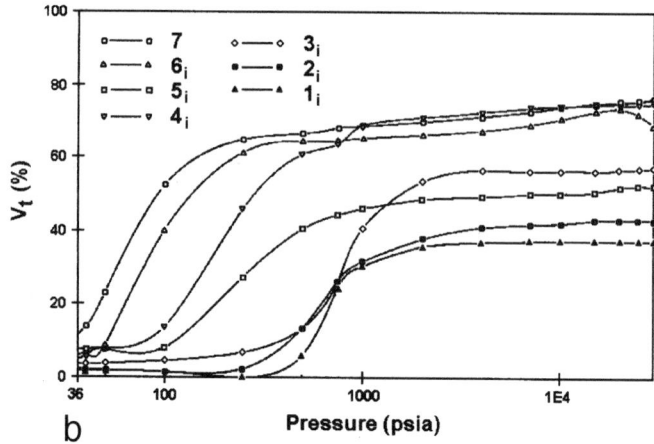

Figure 15.   Continued

where A_1, A_2, and A_3 are the coefficients of the independent variables V_1, V_2, and V_3, which are the volumes of pores of radii greater than 5, 0.5 to 5, and less than 0.5 μm, and DF are the durability factors assigned to selected strata. For the Sphinx limestone PH, 7, and 1_i (Table 9.4), then, one can write the following equations:

$$A_1 \times 49.68 + A_2 \times 24.65 + A_3 \times 25.66 = 10 \tag{9.3}$$

$$A_1 \times 74.68 + A_2 \times 9.09 + A_3 \times 16.23 = 9.5 \tag{9.4}$$

$$A_1 \times 40.69 + A_2 \times 5.5 + A_3 \times 53.81 = 1 \tag{9.5}$$

Solving these simultaneous equations (Appendix C), the values of the coefficients A_1, A_2, and A_3 are obtained as 0.114683, 0.274703, and −0.09621. These coefficients, when used for other strata of the Sphinx, yield durability factors shown in Table 9.5, suggesting a good correlation between the observed and predicted durability factors.

The selection of different pore-size ranges for determining durability factors is evident from Figure 4.7, which shows the sequence of filling of small and large pores when brought in contact with water, or a salt solution. It is shown there that the first to fill are small pores, which then supply water to the large pores. Thus, depending on the availability of water, small pores are likely to be filled more frequently than large pores. Furthermore, it is shown (Fig. 9.14) that the pressure generated in beds with a large volume of small pores are much greater than in those with a large volume of large pores.

TABLE 9.5 Durability Factors (DF) Based on a Linear Regression Analysis and the Application of Artificial Neural Networks (ANN)[a]

Beds	DF[a] Linear Regression	DFb ANN
7	9.5[c]	9.5[c]
6_{ii}	9.3	9.3
6_I	9.1	9.0[c]
5_{ii}	9.0	9.4
5_I	8.3	7.2
4_{ii}	8.0	3.6
4_I	7.0	1.7
3_{ii}	7.4	2.7
3_I	3.8	0.9
2_{ii}	7.3	2.2
2_I	0.9	0.9
1_{ii}	5.2	1.0
1_I	1.0[c]	1.0[c]
PH	10.0[c]	10.0[c]

[a]Source: Punuru, A. R., Chowdhury, A. N., Kulshreshtha, N. P., and Gauri, K. L., Control of porosity on durability of limestone at the Great Sphinx, Egypt, *Environ. Geol. Water Sci.*, **15**(3): 225–232, 1990.
[b]Source: Bandyopadhyay, J. K., Yerrapragada, S. S., and Gauri, K. L., Artificial neural networks and the durability of Sphinx limestone, *J. Mater. in Civil Eng.*, **7**(3): 174–177, 1995.
[c]DF assigned on the observed state of weathering.

The selection of three pore ranges rather than just two, that is, large and small pores, is based on treatment of a large volume of data that shows better correlation when three pore ranges are used. Furthermore, it is apparent from Fig. 9.15a (right side) that there is a hysteresis in the extrusion and reintrusion of small pores of each bed. Thus, a selection of two pore ranges in the small-pore category can be justified.

Finally, let us look at Figure 9.15 to read the various aspects of pore-size distributions discussed in the foregoing paragraphs. First, notice that the largest pore radius shown is 5 μm suggesting that large pores are not directly shown by porosimeter readings; however, as discussed previously, they can be determined from data in Figure 9.15a: cumulative initial intrusion and extrusion (right) and the differential initial intrusion and reintrusion (left). In the first case, the difference in volume between the highest initial intrusion and extrusion shows the volume of large pores. In the second case, the reintrusion curve indicates the volume of small pores for various radii, and the initial intrusion curve indicates the sum of corresponding volume of small and large pores. Notice, however, the actual volumes cannot be read from the curves because the volume data have been normalized with respect to pressure by dividing volume within a pressure range by the mean pressure, but they can be found by using the raw porosimetric data.

We know from Figure 9.15a (left) that specific large pore volumes are connected to small pore radii and their volume. This data has been used to draw Figure 9.15b, which is a plot of such calculated data. It is clear from the comparison of these curves that higher beds, for example bed 7, have a much smaller volume of small pores with which the large pores are connected; therefore, this bed is made of highly durable limestone as shown by durability factors (Table 9.5).

9.4.1.2 *Durability Factors by Artificial Neural Networks*

9.4.1.2 Durability Factors by Artificial Neural Networks The mathematical development of the theory of artificial neural networks (ANN) is given in Appendix A. Here, we will discuss its application in developing durability factors and its advantages over other commonly used regression techniques.

ANN is a sophisticated computational tool that, on the surface, may seem like a mechanism for regression analysis. The routine regression analysis can be performed by many software programs, such as the commonly used Minitab. These programs analyze the data and give a best-fit curve, based on least-square analysis. ANN, however, can be trained to recognize patterns. The common regression analysis may thus be viewed as looking at the data in a planar definition, while ANN, because of its ability to recognize patterns, can view the data in its three-dimensional format. As a result, ANN is able to correlate data more efficiently than other regression techniques. For example, with the pore-distribution data given in Table 9.4, when subjected to a Minitab best-fit analysis, one of the two selected ranges of small pores is ignored, suggesting that a significant correlation with DF can be found by treating the data simply as large (>5 μm) and small (<5 μm) pores. As a result, only an approximate relationship is found between the observed and the predicted durability. Using ANN, however, a more precise correlation can be found according to which the strata of bed 4 and lower are in general less durable than those above bed 4. This follows the petrology of the rocks: Beds 4 and lower, which are shown as less durable by ANN, are sparse biomicrites and the more durable higher beds are packed biomicrites.

In the ANN application to the strata of the Sphinx (Fig. 9.16), the pattern inputs (I) of the selected strata may be thought of as an $n \times p$ matrix, where n is the number of input neurons (such as V_1, V_2, V_3 for each set; Table 9.4), and p is the number of selected data patterns (three in this case; Table 9.5). The input and the target, assigned DF values, form a training pair. The inputs are multiplied by randomly selected weights and these values are squashed, that is, converted to a scale of 0 to 1, that then form input to the hidden layer. The output from the hidden layer is summed. If it varies from the desired target value, the difference error forms the basis for updating the weights. The entire process is repeated until the weights generate values that are in accord with the target value. The details of each operation are given in Appendix A. In Table 9.5, DF by regression analysis and by ANN are given. The ftp site (Appendix C) contains programs in Fortran, by which DF can be determined for any pore-size distributions.

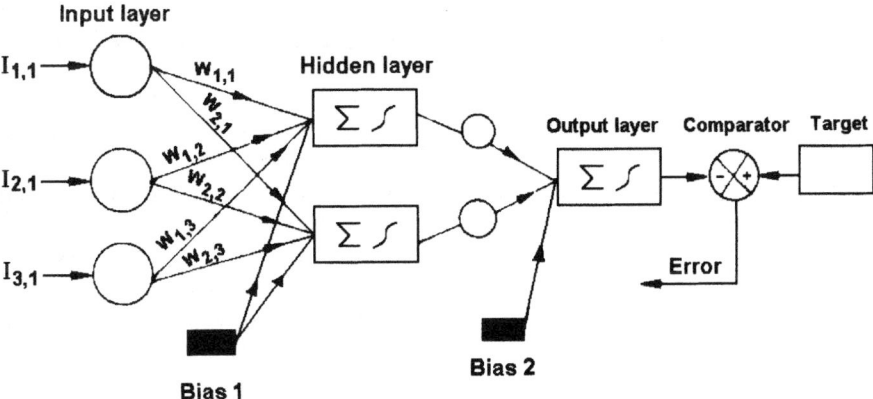

FIGURE 9.16 Artificial neural network (ANN). The ANN architecture is shown for determining durability factors for the Sphinx limestone.

9.4.1.3 Durability Factors by Fractal Dimension

As stated in Chapter 8, pore-size distributions that obey a power law have a fractal structure. A power-law relationship is defined if the regression curve of data when plotted in the form $\log(dV/dP)$ versus $\log(P)$ reveals a definite slope. The gradient of the slope, if noninteger, is called fractal dimension.

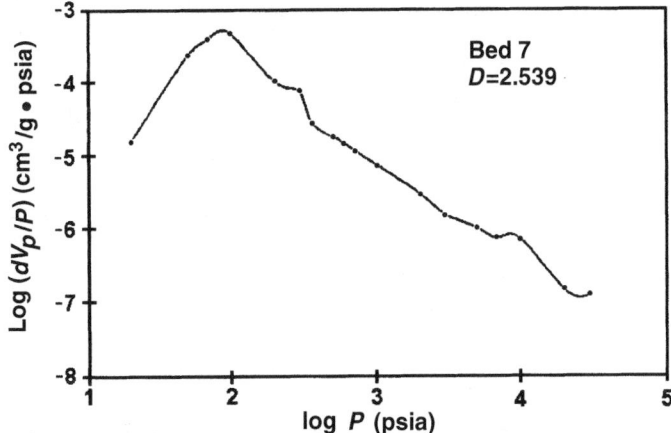

FIGURE 9.17 Power-law behavior of the pore-size distribution. The distribution of pores of diameter less than 1.8 μm (equivalent pressure of 100 psia) describes a slope, which is the fractal dimension (D) of bed 7. Note that the trapped volume of mercury, that is, volumes of large pores, is ignored in this analysis because the initial and reintrusion curves are nearly parallel [Fig. 9.15(a)]. *Source*: Yerrapragada, S. S., Tambe, S. S., and Gauri, K. L., Fractals, pore potential, and limestone durability, in *Rocks for Erosion Control*, C. H. McElroy and D. A. Lienhart (eds.), Special Technical Publication (STP); 1177, Philadelphia: American Society of Testing and Materials, pp. 38–45, 1993. Reprinted with permission, copyright 1993, ASTM.

Any portion of the data of a sample that obeys a power law can be used to determine its fractal dimension. For example, Fig. 9.17 shows the pore-size distributions in bed 7 obtained by mercury intrusion in the range of 15 psia to 30,000 psia. The left portion of the curve does not obey a power law and thus must be excluded from consideration of fractal analysis. However, the upper limit of the fractally behaving portion of pore-size distributions is 1.8 µm (equivalent pressure of 100 psia). In other beds (Fig. 9.18), we find that the upper limit decreases successively so that it is 0.36 µm (360 psia) for the lowermost bed. Figure 9.18 shows the distribution of data that follow power-law behavior for all other strata of the Sphinx thorax and gives their fractal dimension. It should be noted that these slopes are composites of several disjunct slopes of the type shown in Figure 8.8; however, rather than giving a fractal dimension of each section, a generalized trend of the slopes is shown.

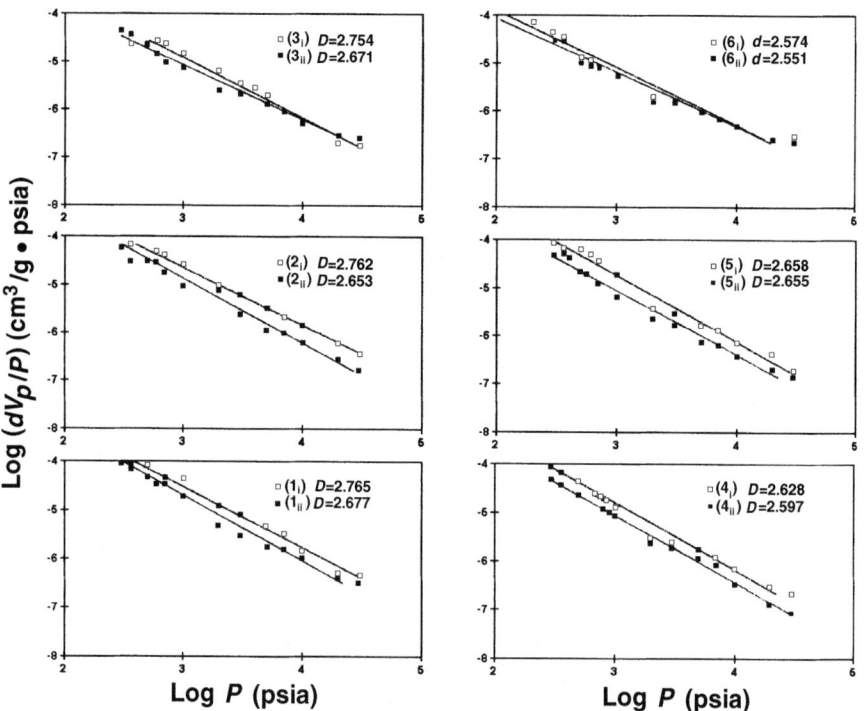

FIGURE 9.18 Fractal dimension of the Sphinx strata showing that the fractally behaving range of pores increases in ascending order of the strata and that each stratum has a unique fractal dimension. The upper curve in each box represents the lower portion (i) of each bed while the lower curve represents the upper portion (ii). *Source*: Yerrapragada, S. S., Tambe, S. S., and Gauri, K. L., Fractals, pore potential, and limestone durability, in *Rocks for Erosion Control*, C. H. McElroy and D.A. Lienhart (eds.), STP 1177, Philadelphia: American Society of Testing and Materials, 1993, pp. 38–45. Reprinted with permission, copyright 1993, ASTM.

TABLE 9.6 Data on Pore Volume and Calculated Fractal Dimension and Durability Factors

Beds	V^a	D^b	$DF_2{}^c$
7	19	2.54	9.5^d
6_{ii}	21	2.55	9.3
6_I	23	2.57	8.9
5_{ii}	38	2.65	6.6
5_I	37	2.66	6.2
4_{ii}	24	2.60	8.2
4_I	27	2.63	7.3
3_{ii}	37	2.67	5.8
3_I	42	2.75	2.1
2_{ii}	32	2.65	6.6
2_I	58	2.76	1.5
1_{ii}	38	2.68	5.4
1_I	61	2.77	1.0^d
PH	18	2.50	10.0^d

Source: Yerrapragada, S. S., Tambe, S. S., and Gauri, K. L., Fractals, pore potential, and limestone durability, in *Rocks for Erosion Control*, McElroy and Lienhart (eds.), STP 1177, Philadelphia: ASTM, 1993, pp. 38–45. Reprinted with permission, copyright 1993 American Society of Testing and Materials.

[a]Volume of small pores (< 0. 5 μm) obtained from initial intrusion.
[b]Fractal dimension: slope of the plot $\log(dV/dP)$ versus $\log(P)$.
[c]Durability factors derived from fractal dimension by Eq. (9.6).
[d]DF assigned on the observed state of weathering.

The result of this analysis can be summarized as (1) pore-size distributions in the selected regimes follow a power law, (2) the D values range from 2 to 3 as one would expect for a pore structure that falls between a surface and a three-dimensional form, and (3) D values increase with increasing small-pore (<0.5 μm) volume, expressed in Figure 9.19 as a percentage of total porosity. The latter can be explained by the greater roughness of the surface, shown by small pores, which simulate more closely a three-dimensional form that approximates a nonfractal dimension of 3.

Finally, we look into the correlation of the fractal dimensions and the durability factors of beds Ph, 7, and 1_i. The expression that fits the data best has a quadratic form, which can be given by

$$DF = a_0 + a_1 D + a_2 D^2 \tag{9.6}$$

where D is the fractal dimension and a_0, a_1, and a_2 are the regression coefficients that have the calculated values $a_0 = -90.58$, $a_1 = 444.022$, and $a_2 = -533.931$. Table 9.6 shows the pore volume, the fractal dimension, and the durability factors calculated using the preceding equation.

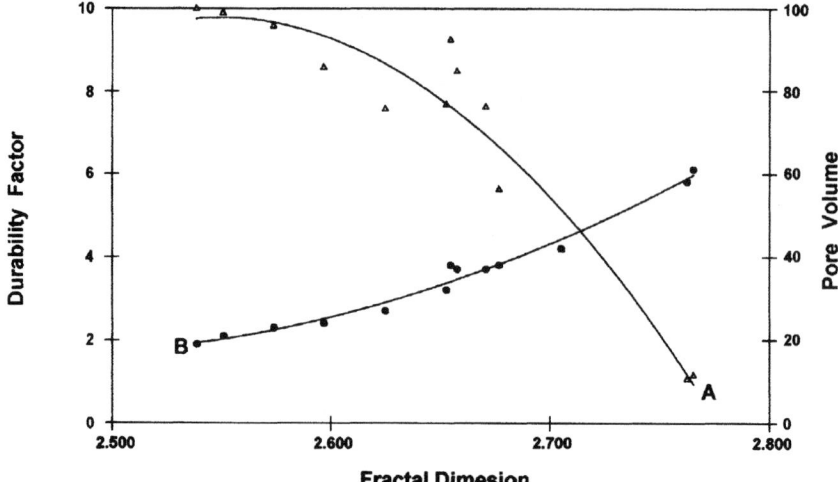

FIGURE 9.19 Correlation of durability factors, fractal dimensions, and pore volumes of the Sphinx strata showing the relationships of the fractal dimensions with DF (curve A) and fractally behaving pore volume (Table 9.6; curve B). *Source*: Yerrapragada, S. S., Tambe, S. S., and Gauri, K. L., Fractals, pore potential, and limestone durability, in *Rocks for Erosion Control*, C. H. McElroy and D. A. Lienhart (eds.), STP 1177, Philadelphia: American Society of Testing and Materials, 1993, pp. 38–45. Reprinted with permission, copyright 1993, ASTM.

The correlation of fractal dimension with durability factors and the pore volume in each layer is shown in Fig. 9.19. Table 8.1 gives selected points from a large array of data obtained by mercury porosimetry. The interpolation of these data by a spline algorithm gives even a larger array. Based upon these data the fractal dimensions were calculated as shown in Table 9.6. The fractal dimensions would be somewhat different, however, if fewer data points were used. The ftp site (Appendix C) has a program that allows determination of DF if the fractal dimension is known.

9.4.1.4 Durability Factors by Pore Potential The pore potential (U), as shown in Chapter 8, is defined as

$$U = W_I - W_E = W_T + W_{HY} \tag{9.7}$$

where W_I and W_E are the *PV* work performed in initial intrusion and extrusion, and W_T and W_{HY} are the works in the entrapment and the hysteresis loop. The pore potential can be related to the durability factors by the expression

$$DF = \frac{W_T}{W_T + W_{HY}} \times 10 \tag{9.8}$$

where the multiplication factor of 10 is used to normalize the scale chosen to be between 1 to 10. Table 9.7 gives these values.

TABLE 9.7 Entrapment Hysterisis (W_T) and Extrusion-Reintrusion-Hysterisis (W_{HY}) Energies, and the Calculated Durability Factors of the Sphinx Strata

Beds	W_T (J/gm)	W_{HY} (J/gm)	DF
7	0.5518	0.0077	9.8
6_{ii}	0.3022	0.0848	7.8
6_I	0.2794	0.1087	7.1
5_{ii}	0.2103	0.1112	6.5
5_I	0.5730	0.2657	6.8
4_{ii}	0.1590	0.0711	6.9
4_I	0.2668	0.3373	4.4
3_{ii}	0.2691	0.1328	6.7
3_I	0.2099	0.2323	4.8
2_{ii}	0.2846	0.2886	5.0
2_I	0.3814	0.9644	2.8
1_{ii}	0.2468	0.3859	3.9
1_I	0.3900	0.7222	3.5

Source: Gauri, K. L. and Yerrapragada, S. S., Pore potential and durability of limestone at the Sphinx, *Mater. Res. Soc. Symp. Proc.*, No. 267, Pittsburgh: Materials Research Society, 1992, pp. 917–923.

Thermodynamic work is a path function. Therefore, dV steps in the calculation of PV work must be very small. As in the case of fractal dimension, the calculated PdV work for durability factors appearing in Table 9.7 was calculated for the entire porosimetric data as well as for interpolation points obtained by the spline algorithm.

9.4.2 Durability and Saturation Coefficient

The saturation coefficient is defined as the ratio between the natural capacity of stone to absorb water by capillarity to its capacity to absorb under pressure. Rocks with a saturation coefficient less than 0.9 are generally considered to be more durable.

To saturate a sample with water, it is first dried at 105°C, cooled in a desiccator, and weighed (W_1). This sample can now be saturated with water by capillary suction by immersing it in water for nearly 24 h and then weighing it (W_2) after it is wiped dry at the surface. ($W_2 - W_1$) is the weight, or volume, of water needed to saturate the sample by capillary suction. We expect that a proportional volume of water will be needed to saturate a wall to a certain depth when exposed to rain.

Now, the sample that we used previously can be reused to fully saturate it, which is done by the application of a vacuum. Figure 9.20 shows a water saturation apparatus that consists of a flask with a thistle funnel connected to a vacuum pump. The dry sample is placed in the flask with the valve of the thistle funnel closed. The thistle

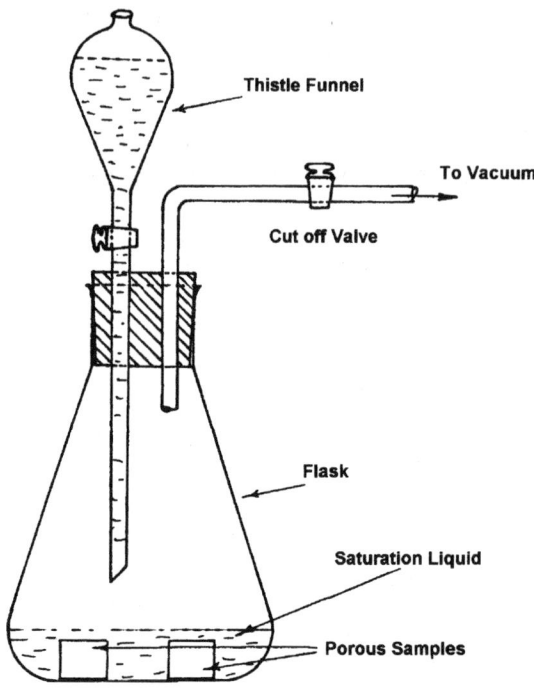

FIGURE 9.20   Apparatus to saturate sample with water using the application of vacuum.

funnel is filled with water while the vacuum is applied on the dry sample. After evacuating the sample for a desired time, the valve of the thistle funnel is opened until the sample is completely covered by water. The vacuum pump is stopped and air is let into the flask. Air pressure on the water surface will force the water into the pores from where the air had previously been drawn out, which facilitates complete saturation. After nearly an hour the sample is taken out of the funnel, its surface is wiped dry, and the sample is weighed (W_3). Now, the ratio $(W_2 - W_1)/(W_3 - W_1)$ is the saturation coefficient.

The difference in the volumes of water that can enter a rock naturally, that is by capillary suction, and that can be forced into it is due to the nature of the pore structure. When a rock with a pore system consisting of many uniform-size capillaries is brought in contact with water, it is likely to become fully filled, more readily filled than is a rock containing many large pores connected to each other by fewer capillaries. The latter, however, will also become fully filled if pressure is applied on the water surrounding the sample. The result is that less water enters the pores than needed to saturate it if many large pores are present and are connected to each other by fewer capillaries.

In cold climates the effect of the saturation coefficient on durability is related to the freezing of water. The idea is that on freezing water experiences a 10% increase in volume. Thus the stone with a saturation coefficient larger than 0.9 will experience

TABLE 9.8   Saturation Coefficients of the Rocks of the Sphinx Thorax

Bed	Saturation Coefficient
7	0.83
6_{ii}	0.83
6_I	0.88
5_{ii}	0.91
5_I	0.94
4_{ii}	0.86
4_i	0.86
3_{ii}	0.96
3_i	0.98
2_{ii}	0.95
2_i	0.97
1_{ii}	0.97
1_i	[a]

[a]Samples disintegrate when immersed in water.

bursting pressure. However, in a rock with a saturation coefficient less than 0.9, the unfrozen water can migrate into the unfilled pores as freezing progresses, relieving the rock of the bursting pressure.

In the case of the Sphinx in which freezing does not occur, the saturation coefficient can relate to salt crystallization. Rocks with a lower saturation coefficient will be affected to a lesser extent because water in completely filled pores can migrate into the empty pore space on crystal growth. Table 9.8 gives saturation coefficients of the rocks of the Sphinx thorax, indicating a close match between durability factors, shown earlier, and the saturation coefficients.

SUMMARY

Sedimentation, stratigraphy, and structural geology of the rocks from which the Sphinx is carved is described, and the influence of the composition, porosity, and joints on durability is delineated.

The profile of the Sphinx thorax reveals a differential weathering of the various layers, controlled mainly by pore-size distributions. Empirical durability factors were assigned to certain rock units, the relative durability of which could be judged by observation. These durability factors were correlated with pore-size distributions by simple regression, by application of artificial neural networks, and by treating the pressure–volume data as thermodynamic work and as a fractal dimension. All these models predict to a varying degree of accuracy the durability of the other strata. The developed durability factors can be used to select limestone for the restoration of the Sphinx and for other applications of porous carbonate rocks in outdoor construction.

CHAPTER 10

Conservation of Carbonate Structures

10.1 INTRODUCTION

The conservation of weathered carbonate-rock structures includes the diagnosis of the problem, establishment of criteria for treatment, the treatment itself that attempts to bring the object back to its original condition, preventive measures to arrest possible future decay, and restoration of fragmented portions. These subjects involve methods and materials common to the disciplines of engineering and physical sciences, which of late have been consolidated into the science and technology of stone conservation. This chapter describes general aspects and materials of stone conservation and gives case histories as guides to the development and execution of conservation projects.

10.2 DIAGNOSIS

A scientific approach to the diagnosis of the problem begins with field investigation in which the visible symptoms of the problem are identified. Then, samples are collected from deteriorated areas along with some sound material from deeper layers or adjacent sections of the building. These samples are then analyzed for chemical, physical, and engineering properties to ascertain changes to the sound stone. The analysis reveals the causes of deterioration and guides the development of specifications for treatment.

It is desirable to have a large volume of sample from the weathered structure. Cores are drilled from the stone; often, however, it is not possible to do so. In this case any kind of sample that can be obtained, for example, fallen pieces and scrapings, is collected. Often, samples are also obtained from quarries that supplied the original material.

Two major maladies afflict stone. As has been described earlier, areas protected from rain have black crusts, while the exposed areas suffer from surface reduction. Another problem, common to both exposed and protected areas, is the structural failure of the rock. Thus, samples are obtained from different areas of the building representative of these conditions.

In the case of rocks with crust, it is important to know the composition of the crust and its thickness, which can be determined by sectioning the rock samples perpendicular to the crust. Often it is necessary to consolidate the rock slightly before sectioning so that the crust does not fall off in the process of preparation (consolidation techniques will be described in the treatment section of this chapter). The sectioned surface is then polished and viewed, preferably by scanning electron microscopy. The elemental composition of this surface can be obtained by the energy dispersive (EDS) microanalytical technique. The mineralogical composition of the crust can be determined by X-ray diffraction (XRD). However, if only scrapings are available, they may be analyzed by leaching the water-soluble minerals in deionized water and determining the composition of water by ion chromatography and then determining the composition of the residue, composed often of soot and quartz particles and heavy metals, by SEM and EDS techniques.

The samples from unsheltered areas are analyzed for efflorescence in the same manner as scarping described previously. A micrographic record of sections of such samples is also made to determine the degree of loss of cement. Furthermore, mechanical properties of these samples such as hardness and strength are measured by techniques given in Chapter 3.

10.3 PERFORMANCE CRITERIA

Performance criteria, or specifications, are the expectations of a treatment. They often pertain to those properties that can be determined quantitatively, for example, the level to which an efflorescence must be reduced or the strength that a consolidated rock must achieve. The specifications are based upon the results of the best treatment in laboratory and small-scale field experiments, the latter being a test performed upon a small area of the structure. Thereby the practicability of proposed treatments can be tested and early results of the treatment judged after a short period of exposure.

The following points are considered crucial for the development of performance criteria and subsequent selection of methods and materials for conservation.

- The treatment should be reversible, that is, it can be removed. Implied by reversibility is that if in the future a better system becomes available, then it should be possible to replace the applied treatment with the new. Reversibility is often an unachievable task. Nonetheless, every effort should be made to apply such materials that have the possibility of removal without the application of highly invasive techniques.

- The consolidant should reach the interior of stone deeper than the zone of weathering to ensure a firm bond between this and the underlying stone (Fig. 10.1). Furthermore, the density of the consolidant should decrease with depth so that a strong interface between treated and untreated zones is not developed.

- The treated stone should largely maintain its permeability, or "breathing" capacity, that is, the surface is not sealed and pores are not completely plugged to allow the release of moisture that is often trapped inside.

- The material of treatment should not absorb atmospheric gases (Figs. 10.2 and 10.3) and ultraviolet (UV) radiation (Fig. 10.4). The decay rate of the treated stone will increase if the preservative can absorb these gases. If the material absorbs UV, then the material is likely to breakdown easily.
- The selected material should be such that a maximum enhancement of the desired property of stone is achieved with the minimum amount of the material used (Fig. 10.5).
- If a large volume of the preservative must be used, for example for restoration, the product should be nonflammable and should have the properties of water absorption, thermal expansion, etc., closer to that of the original stone.
- The requirements for application, for example, mixing materials and observation of safety precautions, should be minimal.

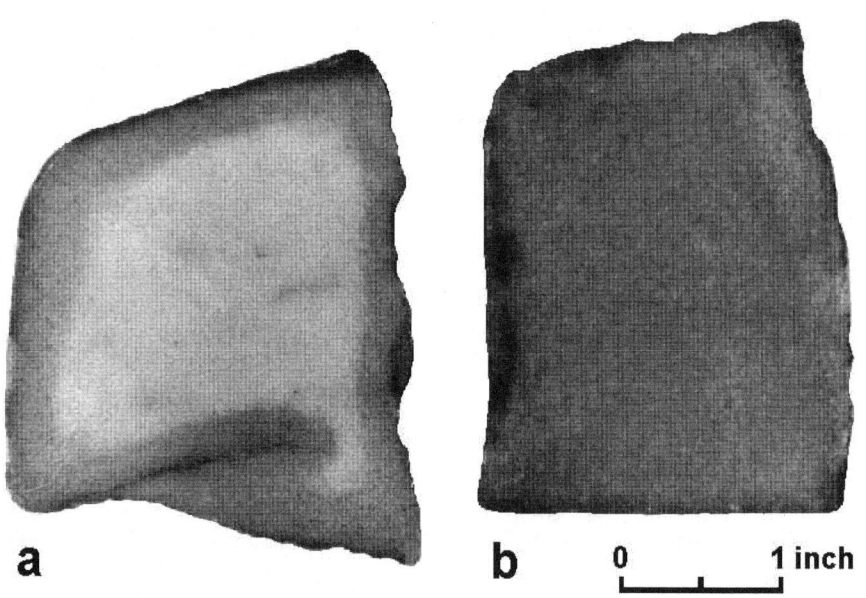

a　　　　**b**　　0　　　　　1 inch

FIGURE 10.1 In-depth impregnation of a weathered marble. Consolidation treatment requires in-depth impregnation. A sample of (a) was treated with an epoxy diluted in a solvent, and a sample of (b) was treated with several solutions in which the concentration of epoxy was successively increased. After treatment the samples were cut in the middle and heated at nearly 200°C. The charred epoxy reveals the depth of impregnation; the treatment such as that given to (a) is undesirable because of the shallow depth of impregnation and the abrupt interface between the impregnated and the underlying stone.

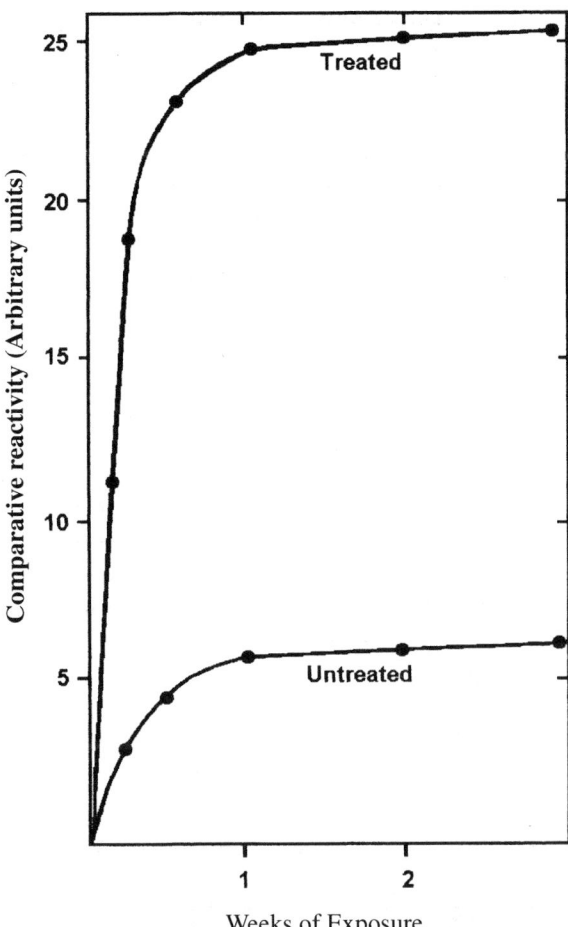

Weeks of Exposure

FIGURE 10.2 Sulfur dioxide absorption by polymers. Sulfur dioxide absorption is indicated by the increased reactivity of marble when treated with an aliphatic epoxy resin. Other experiments showed that this resin absorbed sulfur dioxide nearly one-fourth its unit mass. *Source*: Gauri, K. L., Efficiency of epoxy resins as stone preservatives, *Studies Conserv.*, **19:** 100–101, 1974. Reprinted with permission, copyright 1974, The International Institute of Conservation of Historic and Artistic Works.

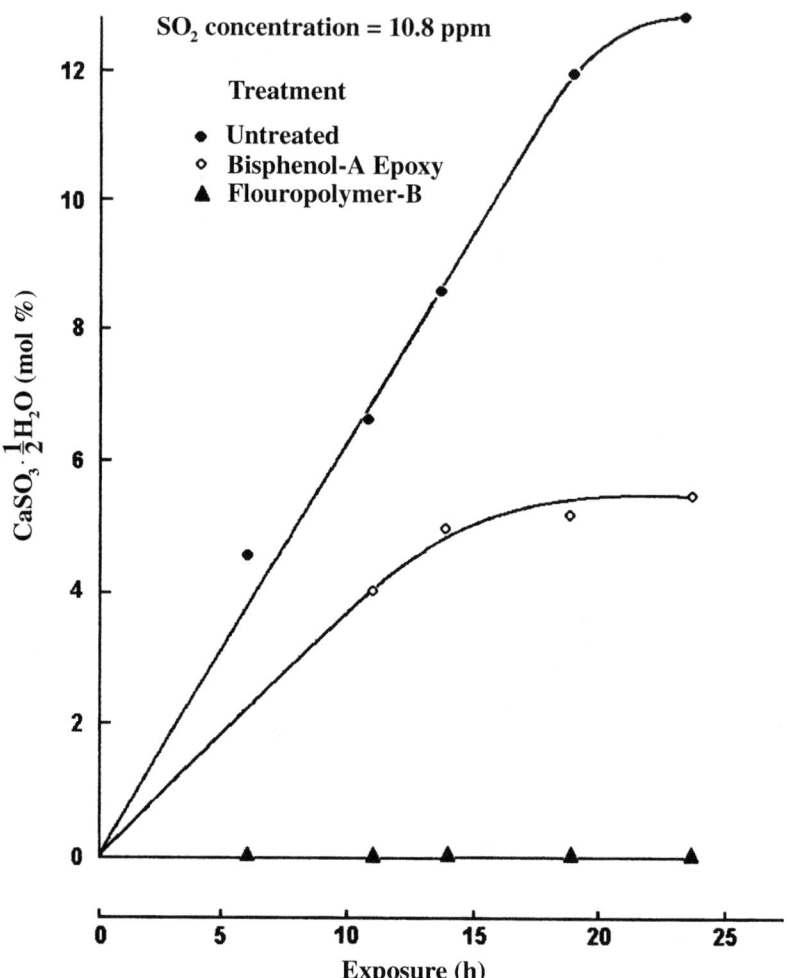

FIGURE 10.3 Relative absorption of sulfur dioxide by polymers indicated by the reactivity of marble samples treated with various polymers. The calcium sulfite hemihydrate reaction product produced is negligible in the sample treated with fluorocarbon as compared with a bisphenol-A epoxy. Other experiments showed that fluorocarbon did not absorb SO_2 while bisphenol-A epoxy absorbed only small quantity. *Source*: Gauri, K. L., Gwinn, J. A. and Popli, R. K., Performance criteria for stone treatment, in 2nd International Symposium on the Deterioration of Building Stones, N. Belloyannis (ed.), Athens, 1976, pp. 143–152.

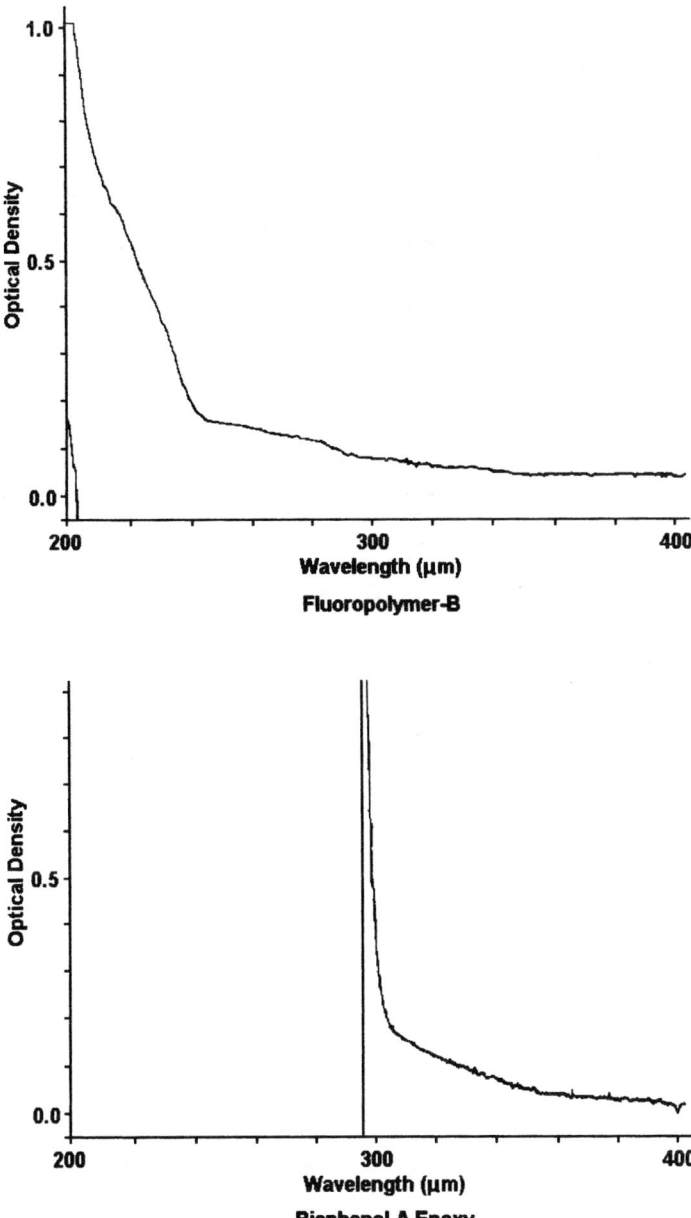

FIGURE 10.4   Absorption spectra of ultraviolet radiation by polymers, showing that epoxy resins absorb a large amount of UV radiation. They are thus highly degradable when present on stone surfaces. Inside the stone protected from UV they act as excellent consolidation materials. Fluoropolymer-B is the same compound as fluorocarbon in Fig. 10.3. *Source*: Gauri, K. L., Gwinn, J. A. and Popli, R. K., Performance criteria for stone treatment, in 2nd International Symposium on the Deterioration of Building Stones, N. Belloyannis (ed.), Athens, 1976, pp. 143–152.

FIGURE 10.5 Consolidation with epoxy resins. A micrograph of a polished surface of a sample of bisphenol-A epoxy impregnated weathered Georgia marble is shown, revealing that a small amount of epoxy resin is able to provide high consolidation. Small plugs of epoxy strengthen the marble so it can be cut and polished. The epoxy is seen as white plugs bridging pores. Stone consolidated in this manner preserves its permeability. *Source*: Gauri, K. L., The preservation of stone, *Sci. Am.*, **238**(6): 126–136, 1978.

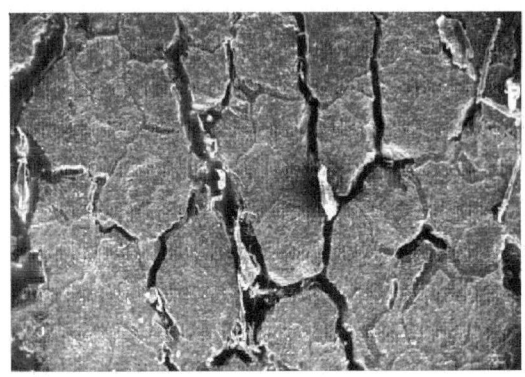

10.4 CONSERVATION TREATMENT METHODS

Historic structures, particularly those in the industrial world, often look dark and gloomy because of the accumulation of street grime and the reaction products of stone and pollutants. Streaks of oxidized metal frequently discolor crumbling stone. Conservation treatments are designed to largely preserve the original characteristics of the structure; they consist of cleaning, consolidating, protecting, and restoring.

10.4.1 Cleaning

Cleaning, as a part of conservation treatment, may be considered under two titles, namely, cleaning crust and other surface accumulations and removing salts from within the stone. We will discuss these in the following but will revisit the subject of salt within the stone in the section on case histories in this chapter.

Most objects protected from rain are dirty due to the accumulation of road dust and combustion residues of fossil fuels, which become embedded in the reaction products of the pollutant gases and the calcareous stone. Efflorescences are often present in porous stone. When present, they disfigure the stone as well as cause damage by repeated solution and crystallization. The microbial films form unsightly encrustation and contribute towards the decay of stone. The deterioration of metallic objects blemish the stone surface while also producing damaging oxidation products. The object of cleaning is to remove these products so that the stone regains its original appearance. However, a balance must be sought between the revival of the appearance and the damage that may be caused by cleaning operation.

10.4.1.1 Cleaning Surfaces
The methods of surface cleaning are chemical, biochemical, mechanical, or combinations thereof. The materials often used for

chemical cleaning are alkalies and acids. Among alkalies, sodium hydroxide (NaOH) is the most commonly used, whereas among acids muriatic acid (HCl), hydrofluoric acid (HF), and ammonium bifluoride (NH_4F_2) are often used. The latter is used in the form of paste in absorbent powders such as bentonite and starch. Certain chelating agents that form water-soluble complexes with metallic ions are also used along with surface-active agents. It seems to be a matter of common agreement that most acids, alkalis, and derivatives thereof remove some original stone. Additionally, they are prone to produce damaging salts in masonry. It is preferred, therefore, that carbonate masonry be cleaned with nonchemical media such as air, water, steam, and solvents. In most cleaning practices mentioned, effort is made that the least amount of water, least water pressure, and least air pressure possible is used.

Organic growths are commonly removed along with other cleaning operations. For specific purposes, however, organometallic salts of zinc and copper are often used.

The cleaning of crust requires a special strategy because there are no known commercially available techniques that have proven to be successful. We will first review the nature of crusts and then give possible means for their treatment.

Crust is an accumulation on the stone surface. The material of the crust is largely derived from the underlying rock. In some ways crusts are similar to the patina upon metals that protect the object from further oxidation. Crust on carbonate rock, however, can be both protective or deleterious, depending upon its makeup and the climate.

Carbonate Crust In Chapter 4 we described the gypsum duricrusts that form in desert climate. They are, like efflorescence, made of soluble salts brought from within to the stone surface, where they are preserved due to the water scarcity in the desert. We also discussed case hardening, which is mechanistically similar to the formation of duricrust; however, now the mineral of the crust is often calcite, which has a very low solubility in water. These recrystallized calcite crusts, as we have shown, can form in any climate, wet or dry. The gypsum and calcite crusts are often clean but if dirty they can be safely removed by the careful mechanical cleaning processes described previously.

Gypsum Crust The problem crusts, however, are those that form by the attack of pollutant gases upon carbonate rock. Such crusts are largely made of gypsum. Embedded in such crusts are particles of soot, coal tar, and silica that cement it tightly to the underlying stone. Water washing, unless with high pressure, is unable to clean such crust. Furthermore, these crusts are often very thin and therefore any mechanical or chemical application of abrasives or chemicals is likely to remove a portion of the underlying rock. In the following we will describe a procedure, based on the application of bacteria *Desulphovibrio desulfuricans*, which has been applied successfully on an experimental scale in cleaning the crust upon marble as well as converting some of the gypsum to calcite.

The background for treatment with *Desulphovibrio desulfuricans* lies in the observation that in Culberson County of west Texas more than 100 hills of limestone rise 3 m to 30 m above the Gypsum Plain. This limestone possesses such features as lamination, microfolds, brecciation, and crystalloblastic texture. These are the characteristics of the underlying rock, which is made of anhydrite ($CaSO_4$). Since the source of this calcite was unknown to geologists, they wondered if the calcite had chemically replaced the anhydrite while preserving the structure of the parent rock.

The anhydrite and gypsum ($CaSO_4 \cdot 2H_2O$) deposits in the Gulf Coast are associated with petroleum fields. It was suggested that the chemoautotrophic sulfur-reducing bacteria *Desulphovibrio desulphuricans* (Chap. 7), abundantly present in petroleum, dissociated the sulfate from anhydrite and provided the organic carbon for conversion of anhydrite into calcite.

Since culture of *D. desuphuricans* is commercially available and media for their optimal growth in the laboratory (Postgate, 1984) are known, it was therefore possible to simulate artificially the natural phenomenon of gypsum calcification through following experiments.

Culture of *D. desulphuricans*, designated as ATCC 29577 by The American Type Collections, Rockville, MD, was obtained. Pure cultures were grown in a Postgate Medium C (5), consisting of

KH_2PO_4, 0.5 g; NH_4Cl, 1.0 g; $CaCl_2 \cdot 6H_2O$, 0.06 g; $MgSO_4 \cdot 7H_2O$, 0.06 g; sodium lactate, 6.0 g; yeast extract, 1.0 g; $FeSO_4 \cdot 7H_2O$, 4 mg; sodium citrate $\cdot 2H_2O$, 0.3 g; sodium ascorbate, 1.0 g, and sodium thioglycollate, 0.1 g. These were dissolved in 1 l water. The iron sulfate concentration was reduced to 0.014 mM because in the process of bacterial metabolic activity iron sulfide (FeS), a black precipitate, had formed. The medium was modified further by adding L-cysteine, 0.1 g. Culture of *D. desulphurican* containing approximately 10^8 cells per ml were incubated for three to four days at 30°C in anaerobic vessels with atmosphere containing 5% CO_2, 10% H_2, and 85% N_2.

An additional sulfate-free medium C was prepared to which Oxyrase EC-100 was added to remove O_2. To this, then, the *D. desulphuricans* culture described above, centrifuged at 5000 × g, and resuspended in an equal volume of sulfate-free medium C was added shortly before the experiment.

Laboratory Experiments Small slabs of Georgia marble that had a nearly 100-μm-thick black crust were used as test samples. Some of the slabs were immersed in broth that was not inoculated with bacteria, and others were immersed in one to which bacterial culture had been added. Both broths contained no sulfate; the idea was that the bacteria would attack and consume gypsum from the stone surface for the sulfate necessary for their survival.

The purpose of these experiments was twofold: one, to see whether the broth without bacteria would have any effect upon the crust; two, to determine whether or not the bacteria could grow by utilizing gypsum present at the stone surface, and whether any conversion of gypsum into calcite would occur. The growth of bacteria was checked by treating some of the samples that had been in bacteria-inoculated broth with gluteraldehyde, which killed the bacteria instantaneously and froze them upon stone in their living position. Whether or not any calcification of gypsum in the crust had occurred was determined by X-ray diffraction analysis (XRD). All the samples subjected to above experiments were observed under a scanning electron

 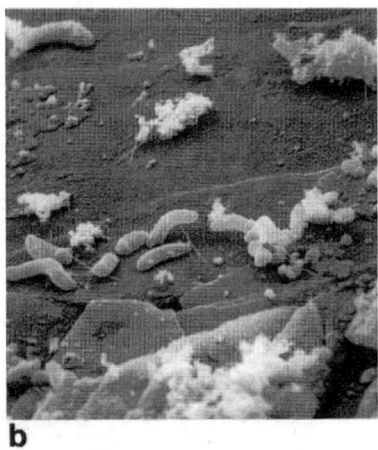

a b

FIGURE 10.6 Cleaning gypsum crust with *Desulphovibrio desulfuricans*. (a) shows a sample that was immersed in the nutrient broth to which the bacterial culture was not added. Notice flocculation of gypsum on the surface and the patches of black crust visible underneath. (b) is a sample that was immersed in inoculated broth and was fixed with gluteraldehyde immediately following treatment with bacteria. The elliptical bacteria seen in the figure have largely consumed the gypsum, with some remaining blobs. *Source*: Atlas, R. M., Chowdhury, A. N., and Gauri, K. L., Microbial calcification of gypsum-rock and sulfated marble, *Studies Conserv.*, **33**: 149–153, 1988. Reprinted with permission, copyright 1988, The International Institute of Conservation of Historic and Artistic Works.

microscope [Figs. 10.6(a) and 10.6(b)]. The result of these experiments can be summarized as follows:

1. Exposure to broth without bacteria did not affect the crust. A flocculation of the gypsum occurred [Fig. 10.6(a)] without removal or change in crust color.

2. The microbes consumed the gypsum and released the carbonaceous material, whereby the blackness was completely removed [Fig. 10.6(b)].

3. The surface of the sample worked by the microbes was smooth, that is, without any granularity, similar to the surface of the gypsum crust. XRD analysis of this surface revealed only calcite, suggesting that at least the gypsum at the surface of the crust had, in situ, transformed into calcite.

Field Experiments Field experiments were then carried out on a marble statue with black crust that had formed over a nearly 110-year exposure in the Cave Hill Cemetery in Louisville [Fig. 10.7(a)]. Before the experiments with bacterial culture were started, patches on the statue where the crust had begun to exfoliate were consolidated by an acrylic resin. Then a watertight shroud of polyurethane plastic was built around the statue.

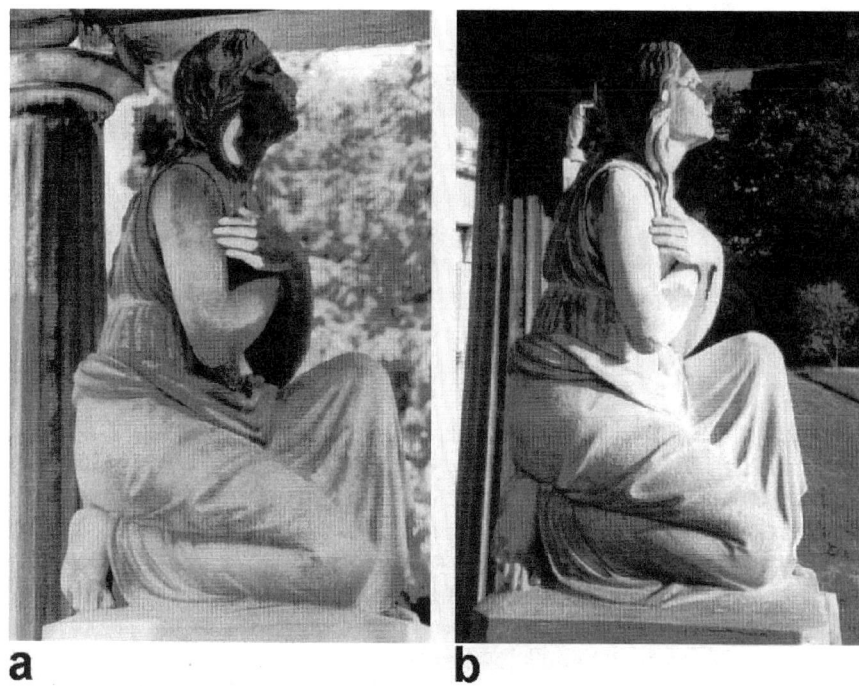

a   b

FIGURE 10.7 Efficiency of bacteria for cleaning crusted marble showing a nearly 110-year-old marble statue in the Cave Hill Cemetery in Louisville (a) before and (b) after treatment with *Desulphovibrio desulfuricans*. Patches on the statue where the crust had exfoliated or was about to do so were consolidated with an acrylic resin before the treatment. Black patches on the nose and arm on the treated statue are the sites where the bacteria could not reach gypsum due to the acrylic coating. *Source*: Gauri, K. L., Parks, L., Jaynes, J. and Atlas, R., Removal of sulphated-crust from marble using sulphate-reducing bacteria, *Stone Cleaning*, G. M. Webster (ed.), London: Donhead, 1992, pp. 160–165.

To test the ability of pure water and the nutrient medium to clean the crust, the statute was first submerged in water and then in the nutrient medium for four days each. In both cases the statue was found to be equally black before and after the treatment.

Then, the statue was submerged in the sulfate-free nutrient medium to which the microbe culture had been added. The culture was allowed to incubate for four days after which the solution was siphoned off. The statue was then lightly rinsed with water. The result is seen in Fig. 10.7(b), showing that the microbial treatment removed the surface blackness, apparently in the same manner as in the case of laboratory experiment. XRD of a scraping from the cleaned surface showed that it was mainly made of calcite with a minor quantity of gypsum, confirming the observation made on the laboratory experiment.

These experiments suggest that through reactions activated by enzymes, the bacterial activity dissociated gypsum into Ca^{2+} and SO_4^{2+} ions. The bacterium then reduced sulfate while the calcium ions reacted with carbon dioxide (CO_2) produced as the bacteria utilized organic lactate in the broth, forming calcite ($CaCO_3$) as

$$6CaSO_4 + 4H_2O + 6CO_2 \rightarrow 6CaCO_3 + 4H_2S + 2S + 11O_2$$

Thus, not only was the black crust cleaned but at least some of it had recrystallized back to the parent calcite. This result has an implication to the future weathering of cleaned stone.

We now know that the crust is an external appendage, covering underneath the original surface of the statue. It seems that at least the outer layers of the crust can be partly changed to calcite. This new calcite layer can act as a sacrificial membrane for future reaction with the polluted atmosphere. Knowing the rate of formation of crust in a given atmosphere (Chap. 6) a periodic treatment can be developed.

10.4.1.2 *Cleaning Efflorescences* Efflorescences are present at and below the surface of stone. They may be removed from stone surfaces by brushing, but those lodged in the pores are difficult to remove. The common method of cleaning efflorescence from depth is by clay poultices. The wet poultice, often made of attapulgite—a clay mineral (Chap. 2)—is applied to the stone surface. This tends to pull the salt into the poultice. When dry, the poultice is replaced with a new wet poultice and the process repeated until the salt is reduced to an acceptable level. We will later describe in case histories two additional methods of removal of efflorescence.

10.4.2 Consolidation

Consolidation is the strengthening of the rock mass by introduction of cement in the pores. Consolidation is often necessary for rocks exposed to rain and for sheltered rocks in which the crust has exfoliated. Consolidation is often achieved by in-depth impregnation of rock with organic adhesives, popularly called resins. The resins are used for their ability to dissolve in organic solvents. As a result, their viscosity can be reduced to a desired level to enable deep penetration into the pores (Fig. 10.1).

Adhesives commonly used for consolidation are monomers and polymers of acrylics, epoxies, polyesters, silicones, urethanes, and vinyls. Epoxies, among these resins, for example, are thermosetting, that is, they set, or become solid, by the addition of curing agents that cross-link with the resin and transform the resin into an insoluble mass that can bond grains together. The acrylic polymers are thermoplastics; they solidify from a solution as the solvent evaporates. In each case, the solid polymeric material provides cohesion between the grains (Fig. 10.5). However, the strength provided by thermosetting resins is generally larger than that of thermoplastics. Thus, a nearly comparable degree of consolidation can be achieved by a relatively smaller quantity of thermosetting resin than a thermoplastic resin. We will describe in detail later the use of an epoxy resin in the conservation of the California Building. Recent studies of resins used in historic preservation are prepared, among others, by Charles Selwitz of Getty's Conservation Institute, who was kind to write the preface to this book. His work on epoxy resins (Selwitz, 1992) is particularly noteworthy and another on acrylics is in preparation (personal communication).

Some inorganic materials, given in the following section on surface protection, have also been used as consolidants for carbonate rocks.

10.4.3 Protection

After cleaning and consolidation, clean surfaces as well as those of newly installed objects of carbonate rocks expose fresh calcite to the atmosphere. Since SO_2 is the most aggressive gas in industrial environments, we will discuss the protection from this gas, noting that such a protection will also be effective against other chemically active gases such as CO_2 and NO_2. Commonly proposed measures to protect monuments from SO_2 attack include changes to the surface chemistry and the applications of coatings.

10.4.3.1 Change to Surface Chemistry The extent of chemical reaction of a stone in an SO_2 environment is controlled by the buffering capacity of the constituent minerals. The carbonate minerals have a large capacity to buffer acids and acidic gases. Therefore they are highly reacted in an SO_2 environment. A reduction in this capacity by changing the mineral composition will reduce the extent and the rate of reaction. We give examples of some anionic and cationic surfactants that replace CO_3^{2-} or Ca^{2+} so that the new mineral is attacked by SO_2 to a lesser extent. The discussion that follows is based upon recent studies of Hasan Boke at the University of Louisville.

Some of the common anionic surfactants are phosphates, oxalates, and oleates. Their soluble species such as dipotassium monohydrogen phosphate (K_2HPO_4), potassium oxalate ($K_2C_2O_4 \cdot H_2O$), and sodium oleate ($C_{17}H_{33}COONa$) are commonly used. The reaction of calcite with these solutions can be expressed as follows:

$$CaCO_3 \rightarrow Ca^{2+} + CO_3^{2-}$$

$$Ca^{2+} + C_2O_4^{2-} \rightarrow CaC_2O_4 \cdot H_2O$$

$$Ca^{2+} + HPO_4^{-2} \rightarrow CaHPO_4 \cdot 2H_2O$$

$$Ca^{2-} + 2C_{17}H_{33}COO^- \rightarrow Ca(C_{17}H_{33}COO)$$

The newly formed compounds at the stone surface, namely, calcium oxalate, calcium phosphate, and calcium oleate, are more resistant to SO_2 attack than calcite.

Cationic substitutions commonly refer to the replacement of calcium by barium. In the mid-nineteenth century efforts were made to preserve the decayed limestone of the New Palace at Westminster by using a barium hydroxide [$Ba(OH)_3$] solution to produce an inert compound, witherite ($BaCO_3$), or a solid solution of barytocalcite

$BaCa(CO_3)_2$. Since then many related processes based on the substitution of Ca^{2+} by Ba^+ have been designed.

10.4.3.2 Organic Coatings Synthetic monomers and polymers used as consolidants automatically leave behind surface coatings. These, and those applied specifically as surface treatment, can be protective. Often, however, coatings form an impermeable layer upon stone surface. In the case, they peel off soon after application while having also caused accelerated degradation of stone. Consequently, it is desirable to remove the coating produced by the consolidation treatment. Such surfaces and those that have been freshly installed often perform better if no coatings are applied. Nevertheless, coatings that largely maintain permeability of the stone and are made of non-UV, non-SO_2-absorbing media, such as fluorocarbon (Fig. 10.3), can effectively reduce the surface reactivity of carbonate rock in a polluted environment.

10.4.4   Restoration Materials

When portions of the stone have disintegrated, selection of materials to replace the worn stone is needed. Organic polymeric composites are often used for patching. Primary considerations in the design of such media is that their properties largely match the original stone and that the composites themselves are durable. We will discuss specific properties below in the context of the treatment of the California Building.

Restoration may sometimes require repair of fractures that develop due to the structural load or following an earthquake event. The fractures are often treated by injection with epoxy resins. In practice, the upper surface of the crack is sealed, and then epoxy resin admixed with the curing agent is injected under pressure at one end of the crack until it appears at the other end. After the polymerization of epoxy the sealant is removed and the stone surface cleaned, sometime using abrasive materials.

10.5   CONSERVATION OF THE CALIFORNIA BUILDING: A CASE HISTORY

The California Building in San Diego was constructed in 1915. The major element of the building is the central dome, which is flanked by four vaults and a bell tower. The latter is located on the east side of the building, has ten stories, and is nearly 200 feet high. The building has elaborate ornamentation. The ornamentation at the main entrance of the building and the upper three stories of the tower as well as the stuccoed front facade of the lower seven levels of the tower is the subject of conservation.

The building is made of cast stone, which is not a natural carbonate rock, but was manufactured by cementing quartz sand with calcium carbonate cement. In addition, this stone is porous like limestone and dolomite. Over time the stone had weathered and needed conservation treatment. The building was treated in 1975 after a thorough investigation by us of the causes of decay and the properties of the cast stone.

The weathering had produced several visible features (Fig. 10.8):

FIGURE 10.8 The California Building. Showing efflorescence, erosion, and fragmentation of the cast stone. *Source*: Gauri, K. L., Conservation of the California Building, in International Symposium on Deterioration and protection of Stone Monuments, N. Belloyannis (ed.), Paris: RILEM, 1978, p. 1–14.

1. Reduction of surface relief had occurred, due to dissolution of the calcitic cement, which caused separation of sand grains. The erosion was particularly pronounced at ornamental surfaces.

2. Fragments of stone had fallen off from several sites. They were often associated with the occurrence of efflorescence.

3. Efflorescences were visible as white encrustation. Although present everywhere on the building, they were highly visible as the cause of stone disintegration at inner surfaces of the upper three stories of the tower because of frequent cycles of wet and dry episodes.

4. Iron bars were used to connect the ornamental cast stone to the structural concrete. In many places iron bars had oxidized, as a result of which both the concrete and the cast stone had ruptured.

5. Finally, heavy and poorly supported members of ornamentation were considered unsafe due to earthquake activity in the region. They had to be dismantled. Thus, the added function of the conservation treatment was to design materials for patching and for production of lightweight replicas.

Laboratory tests were performed on samples extracted from the building. Core samples 4 inch in diameter and of variable length were obtained so that the properties of the near-surface weathered zone and the underlying unweathered zone could be studied. Some of the cores were treated (given later) and comparative properties of the treated and untreated samples determined. We will give results of these experiments along with the procedures and specifications for the treatment.

10.5.1 Cleaning

The facade had some street dirt and soot but the main problem was the occurrence of efflorescence. X-ray diffraction showed that the efflorescence consisted of gypsum $(CaSO_4 \cdot 2H_2O)$, glauber salt $(Na_2SO_4 \cdot 10H_2O)$, and complex salts such as $Na_2Ca(SO_4)_2$. The ionic composition was determined by atomic absorption spectrophotometry and by ion chromatography (Table 10.1).

The data permitted comparison of the quantities of efflorescences in different parts of the building as well as a means for developing specification. As often the minimum quantity of Na^{2+} found was to be nearly 0.02% and the stone containing it was not damaged, it was therefore specified that: after cleaning the stone shall have no more than 0.02% Na^{2+}. Interestingly, this specification matched that recommended by the British Standards for the selection of brick. Experiments showed that the efflorescence could be reduced to the specified level by the following sequence of operations.

TABLE 10.1 Ionic Composition (Percent Weight of Dry Sample) of Salts Extracted in Deionized Water

Sample	Na^+	Ca^{2+}	SO_4^{2-}
1^a	0.046	0.009	0.060
$1w^a$	0.022	0.017	0.063
2^a	0.046	0.001	0.008
$2w^a$	0.029	0.077	0.218
3^a	0.020	0.005	0.005
$3w^a$	0.017	0.125	0.300
$4w^a$	0.019	0.390	0.185
5^a	0.041	0.003	0.133
$5w^a$	0.049	0.009	0.354
6^a	0.008	0.075	0.018
$6w^a$	0.005	0.034	0.011
1^b	5.210	0.032	6.420
2^b	0.741	0.030	2.040

Source: Gauri, K. L., Conservation of the California Building, in International Symposium on Deterioration and Protection of Stone Monuments, Paris: RILEM, pp. 1–14.

[a]Core samples from near the base of the building; w refers to weathered portion of the sample.
[b]Peels from inside tower.

First, the stone surface was brushed clean of the visible efflorescence. The stone surface was then washed with a combination of water and sand. This process cleaned the stone of other dirt while also removing some efflorescence. The stone surface was then treated with steam at high pressure followed immediately by washing with cold water. After washing, acetone was dripped on the surface to remove the water that carried the efflorescence. The procedure was repeated until the efflorescence was reduced to the specified level.

The cleaning of efflorescence was a major undertaking. The authorities adhered strictly to the compliance of this specification by the contractor. However, it was noticed sometime after the completion of the project that peeling of stone at a few spots had occurred inside the tower and the salt could be seen to be associated with the peels. Apparently, this area was not fully cleaned as specified.

10.5.2   Consolidation

The consolidation of the core samples, selected after various trials, was by immersion in a solution of 15% bisphenol-A epoxy in acetone for 20 min followed by immersion in 25% epoxy solution for 10 min. Diethylene triamine was the curing agent. A similar treatment of a test area of the facade for a corresponding time period, but applied by brushing, impregnated the stone to a depth of over 1 inch. However, because of quick evaporation of acetone in the field experiment, the recommendation for epoxy concentrations was changed to 10% and 20% for the treatment of the building. The consolidation treatment recovered cohesion between grains and also changed other properties of the stone including compressive strength, water absorption, and permeability, discussed below.

10.5.2.1 Compressive Strength Compressive strength reflects the overall mechanical behavior of stone. The improvement of compressive strength automatically improves the tensile strength as well as the abrasive index. As a result the stone does not easily lose grains and chances of failure due to crystallization of salt are reduced.

The compressive strength of the weathered zone had decreased considerably (Table 10.2). For example, comparison of sample 1 Au (zone of weathering intact) with 1 Bu (zone of weathering cut away) shows that the compressive strength was reduced by nearly 15% due to weathering and that the treatment improved the compressive strength of 1 Au by more than 100%. Similarly, the compressive strength of the unweathered portion of the stone had also increased by the treatment. The overall data (Table 10.2) permitted the development of the specification that:

> The stone shall be impregnated to an average depth of one inch and that the compressive strength of the treated stone shall increase at least 50% of the untreated weathered portion of the stone.

TABLE 10.2   Compressive Strength of Treated and Untreated Stone[a]

Sample	Compressive Strength (kg/cm^2)
1 Au	231
1 At	562
1 Bu	273
1 Bt	382
2 Au (chip missing)	99
2 At	241
2 Bu	166
2 Bt	425
3 Au	295
3 At	398
3 Bu	155
3 Bt	388

Source: Gauri, K. L., Conservation of the California Building, in International Symposium on Deterioration and Protection of Stone Monuments, Paris: RILEM, 1978, pp. 1–14.

[a]The samples were 1.5 inch in diameter and 3 inch-long cores drilled from 4 inch diameter core samples. The compressive strength was measured by a universal testing machine. A and B indicate, respectively, cores with and without the weathered zone and t and u represent samples cored from consolidated and unconsolidated larger cores.

10.5.2.2   *Water Absorption*

Deterioration of stone is directly related to its water absorption. One of the functions of the consolidation treatment is to reduce water absorption and rate of water movement into the stone.

Water absorption was determined by immersing samples in water for 48 h. The consolidation treatment described before reduced the water absorption of weathered samples from nearly 12% to 1.5%. To measure the rate of water absorption, two plane surfaces were cut normal to each other, both being normal to the surface of weathering. One of the plane surfaces was brought in contact with water and the capillary rise of water observed at the plane surface normal to it (Fig. 10.9). It was found that water in one of the treated samples did not rise into the region near the surface of weathering even after 48 h of contact. On the basis of these findings, the following specifications were recommended:

After treatment the weathered stone shall not have more than 5% water absorption and that in the first 24 h of contact with water, the water shall not rise into the near-surface region of the consolidated stone.

TABLE 10.3 Water Absorption (Percent Weight of Untreated Dry Sample)[a]

Sample	Water Absorption
1	2.34
1W[a]	10.04
2	4.87
2W[a]	12.62
3	4.88
3W[a]	7.63
4W[a]	7.75
5	3.25
5W[a]	5.59
6	5.18
6W[a]	7.52
7	11.99
7W[a]	12.64

Source: Gauri, K. L., Conservation of the California Building, in International Symposium on Deterioration and Protection of Stone Monuments, Paris: RILEM, 1978, pp. 1–14.

[a]W indicates specimens from the weathered portion of the stone.

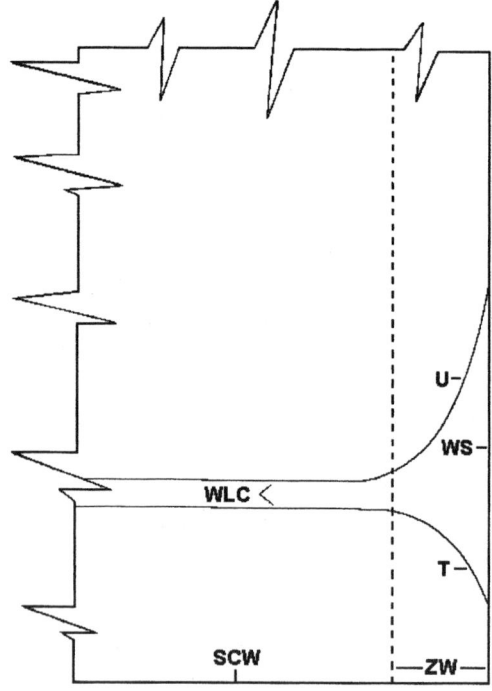

FIGURE 10.9 Capillary rise of water, shown by a schematic presentation. T and U represent profiles of water level in a well-treated and an untreated stone kept in contact with water for an extended period. Other symbols are WS, weathered surface; SCW, surface in contact with water; WLC, water level by capillarity; ZW, zone of weathering. *Source*: Gauri, K. L., Gwinn, J. A. and Popli, R. K., Performance criteria for stone treatment, in 2nd International Symposium on the Deterioration of Building Stones, N. Belloyannis (ed.), Athens, 1976, pp. 143–152.

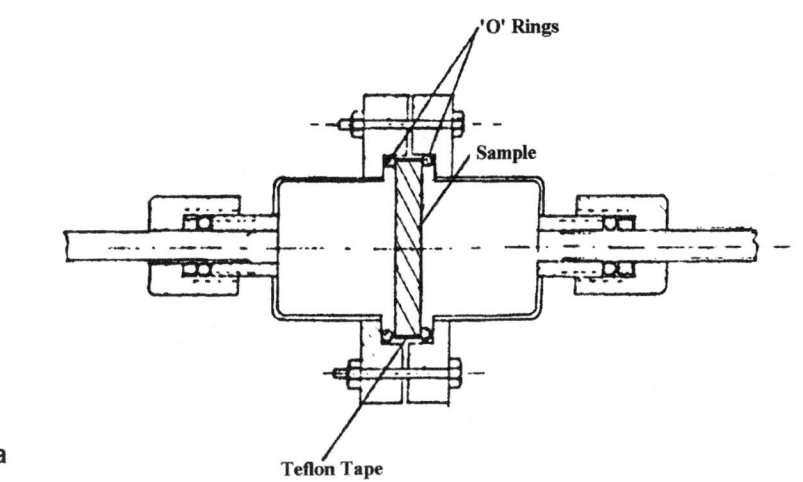

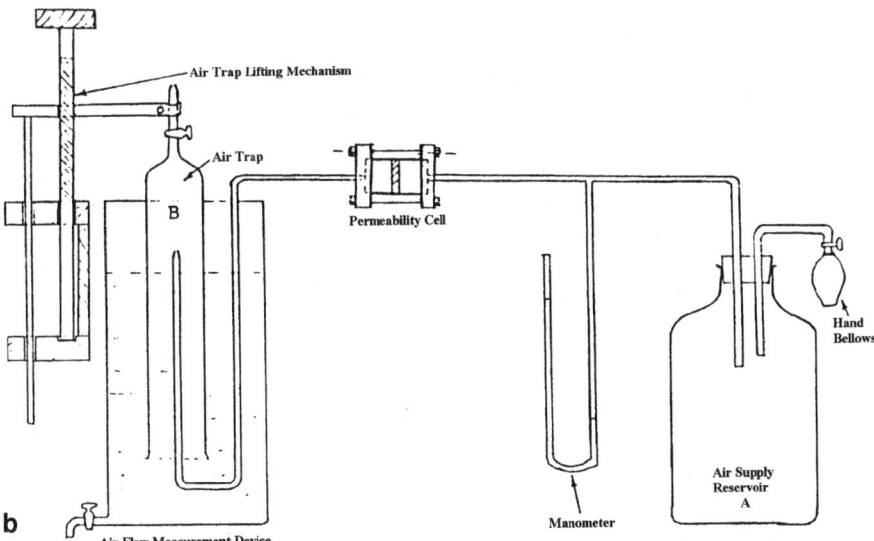

FIGURE 10.10   Measuring the permeability of stone. (a) shows a permeability cell and (b) a schematic diagram of forced-flow measurement apparatus. The manometer measures the pressure of the forced air while the volume of the air passing the sample is measured in the graduated tube, which can be lifted to keep the atmospheric pressure in the trap constant. *Source*: Verma, S. K., Flow and diffusion of gases through porous media, Ph.D. thesis, University of Louisville, 1973, p. 40–41.

10.5.2.3 Permeability Permeability is the ability of a fluid to move through a porous medium. A treated stone should maintain at least some of the permeability so that moisture, if present in the stone, can exit at the surface. To measure permeability, a cell was constructed [Fig. 10.10(a)], which consisted of two steel cylinders, each tapered on one side. The large openings of the cylinders had smooth flanges with holes so that screws passing through holes in the opposite flanges could fasten the two cylinders together into one cell. The sample was placed between the flanges. Air was forced under a known pressure from one side of the sample and collected on the other. The apparatus to measure permeability is shown in Fig. 10.10(b). From permeability measurements following specification was derived:

> A sample taken from the treated facade shall have at least 90% of air flow maintained in the impregnated zone as compared with the flow in the deeper unimpregnated zone.

Finally, after the building had been treated, the surface was lightly sanded to remove the surface epoxy film. If the epoxy film had been allowed to stay, it would have, by absorption of ultraviolet radiation, yellowed over time and eventually fallen off. However, sanding instantly removed the epoxy film. To reduce the effect of air pollutants upon this surface, the entire surface of the building was treated with a 2% to 3% solution of fluoropolymer-B (Figs. 10.3 and 10.4) in a solvent consisting of equal parts of cellusolve acetate and methy-ethyl ketone. The permeability specification was to be met after surface protection treatment.

10.5.3 Restoration Materials

Another major aspect of the treatment design of the California Building was the replacement of the heavy ornamental objects such as lanterns and vases at high elevations with lighter shells molded from synthetic material. Also, the areas from which pieces of stone had fallen were to be patched. In the design of synthetic composite materials for the manufacture of strong, yet lightweight replicas and for patching, it was deemed crucial that the properties were similar to those of the cast stone. These properties were compressive strength, water absorption, permeability, linear expansion, flammability, and color retention.

The manufactured composites consisted of the following components:

1. Inorganic Aggregate. This consisted of fused quartz and natural sands of grain size and color comparable to the grains of the original stone. Whereas the fused quartz helped reduce the coefficient of thermal expansion, the graded particles helped to modify this and other physical properties including color, compressive strength, water absorption, and permeability.
2. Organic Polymers as Cement. The selected polymeric cement consisted of a prepolymer of fluoropolymer-B and an acrylic dissolved in a solvent.

These two components were mixed in varying proportions so that the end product met the specifications given in Table 10.4.

a

FIGURE   10.11   Conservation treatment of the California Building. (a) shows the condition of stone before treatment and (b) after treatment. *Source*: Gauri, K. L., Conservation of the California Building, in International Symposium on Deterioration and Protection of Stone Monuments, N. Belloyannis (ed.), Paris: RILEM, 1978, p. 1–14.

b

TABLE 10.4 Specifications for Composites[a]

Properties	ASTM Method	Values
Compressive strength	C-170	> 200 kg/cm^2
Modulus of rupture	C-99	> 50 kg/cm^2
Water absorption	C97	< 4%
Permeability	C577	± 5 centidarcy
Color retention	G25	100 h exposure, no color change
Coefficient of thermal expansion	E228	< 2.2 × 10^{-6}/°C
Burning characteristics	E162	4–10 (flame index)

Source: Gauri, K. L., Conservation of the California Building, in International Symposium on Deterioration and Protection of Stone Monuments, Paris: RILEM, 1978, pp. 1–14.

[a]The tests (methods) are as recommended by the American Society for Testing and Materials (ASTM) and the values are the specifications developed by us.

Figures 10.11(a) and (b) show a portion of the California building before and after the treatment. An inspection of the building in 1994 showed that the building was in a good condition after the treatment, which was completed in 1975.

10.6 CONSERVATION OF THE SPHINX

At the request of the Egyptian government we once prepared a treatment plan for the conservation of the Sphinx to include (1) cleaning and consolidation of the bedrock, (2) selection of limestone for replacement of worn-out blocks from the constructed veneer, and (3) formulation of mortar for use between those blocks. The selection of limestone for veneer can be based on durability factors discussed in Chapter 9. A stone with DF 9.5 or above should be acceptable (see Table 9.5). In the following we discuss the treatment of the bedrock and the formulation of mortar.

10.6.1 Removing Efflorescence from Bedrock

Bedrock is made of several limestone layers, but has a common problem of efflorescence that has rendered some of the layers highly friable. Thus, the use of a large quantity of water for cleaning is not advisable. Furthermore, because of the loss of cohesiveness between grains due to dissolution of calcite, the stone must be consolidated somewhat before any cleaning operation. A light consolidation was recommended by impregnating the stone with ethyl silicate. This product, on hydrolysis, produces silica (SiO_2), which is able to bond the grains together:

$$Si(OCH_2CH_3)_4 + 4H_2O \rightarrow Si(OH)_4 + 4CH_3CH_2OH$$

$$Si(OH)_4 \rightarrow SiO_2 + 2H_2O$$

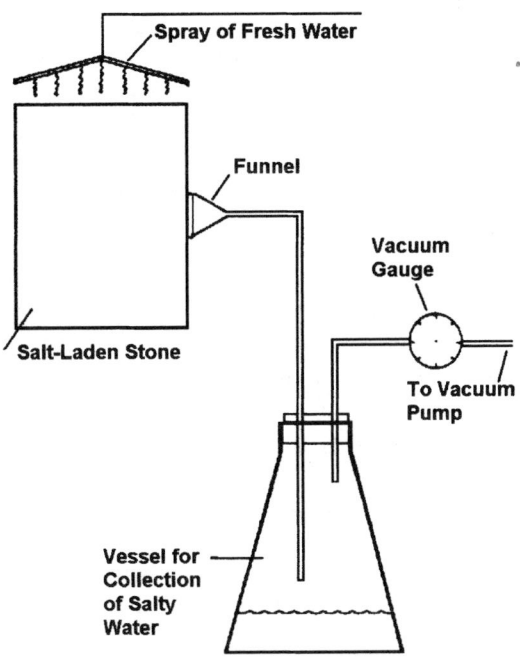

FIGURE 10.12 Vacuum desalination. (a) shows the experimental setup for removal of salt from inside the stone by the application of vacuum. (b) shows two samples: (1) one cleaned by simple water washing and (2) the other by combined water and vacuum application; notice salt encrustation in sample (1), but its absence in sample (2). *Source*: Gauri, K. L., Holdren, G. C., and Vaughan, W. C., Cleaning efflorescences from masonry, STP 935, J. R. Clifton (ed.), Philadelphia: ASTM, 1986, pp. 3–13. Reprinted with permission, copyright 1986, American Society for Testing and Materials.

The desalination was then to occur using vacuum and water by a technique described below. This technique was developed on slabs of Indiana limestone that were abundantly available for experiments.

The slabs were artificially impregnated with the saturated solution of sodium chloride and dried at room temperature. The concentration of sodium was measured in weight percent of stone and found to be 0.12 slightly below the surface, decreasing to 0.08 at a depth of nearly 3 cm from the surface. The greater concentration near the surface was due to drying taking place at the stone surface.

A funnel, connected to a vacuum pump, was attached to the sample surface using a sealant that could be easily removed after the experiment. The stone was sprayed with water so that it was covered by a sheet of water on all sides [Fig. 10.12(a)]. After nearly one-half hour of water spray in which water was able to penetrate nearly 1 inch into the stone, the vacuum pump was started. Analysis by atomic absorption spectrophotometry of the water collected at the end of the vacuum funnel nearly one-half hour after the start of the vacuum pump showed no Na^+, at which point the experiment was stopped and the stone dried. Figure 10.12(b) shows two salt-impregnated samples both continuously washed with water but one (right) to which vacuum was applied. It is apparent that on drying salt did not appear on the vacuum-treated specimen, while significant efflorescence had appeared on the stone surface to which suction had not been applied.

The experiments were conducted using a laboratory vacuum pump with the capacity to displace nearly 150 l of air per minute and a 7.6-cm (3-in.)-diameter funnel as a port. The application of industrial compressors, which can displace a much larger volume of air, and a proper design of batteries of exit ports should permit cleaning of hundreds of square feet of surface per day.

10.6.2 Production of Mortar

Mortar is the masonry component that separates dimensional blocks of a structure. Of several functions that mortar serves, the most important seems to be its ability to drain water out of the building. Thus, by design mortars are porous materials manufactured in the same fashion as the composites described previously.

Old mortars were made by using natural lime as the cementing medium and an appropriate aggregate often made from clean—devoid of salts—river sands of various grades. The lime cement was made by burning limestone at elevated temperature. By this process calcium oxide, called quick lime, is produced, which is then made into a thin paste in water that is slowly churned. The reaction of this paste with atmospheric CO_2 results in the formation of microcrystalline calcite, which is able to bond the grains of aggregate as well as the mortar mass with the surrounding blocks of the natural rock. The reactions taking place are

$$CaCO_3 \xrightarrow{\text{elevated temperature}} CaO + CO_2$$

$$CaO + H_2O \rightarrow Ca(OH)_2$$

$$Ca(OH)_2 + CO_2 \rightarrow CaCO_3 + H_2O$$

In order to formulate new mortar, the composition of the decayed mortar is determined. The following method that we used to analyze old mortars from the Sphinx helped us design the new.

1. Leaching water-soluble salts in deionized water and determining composition of these salts.
2. Digesting calcite in dilute hydrochloric acid in which the constituents of aggregate, which are generally made of clay, siliceous, and iron minerals, are not soluble. By this procedure, the ratio of calcite to aggregate is determined. [We have learned recently from Professor Ettensohn (personal communication) that aggregate sands in mortars often contain large quantities of carbonate sand, the quantity of which can be estimated by petrographic techniques.]
3. Sieving the aggregate to determine ratio of clay (< 0.004 mm), silt (0.004 mm to 0.06 mm), and sand (0.06 mm to 2 mm) fractions.

Thus, when the proportions of cement to aggregate and those of the components of aggregate are known, new mortar, approximating this composition, can be formulated. However it is often found that the modern manufacture of mortar by this technique lacks the mechanical strength necessary to bond the aggregate. The additional strength may be obtained by replacing a small quantity of natural lime with the portland cement. A larger quantity of portland cement in mortar is avoided because this cement has the potential to make less porous, impervious, and hard mortars that can also be chemically detrimental to the stone.

SUMMARY

The conservation of stone structures requires development of specifications for cleaning, consolidation, and protection. Certain methods and materials of conservation treatments are described. The case history of the conservation treatment of the California Building is presented for which the specifications for treatment were developed on the basis of laboratory analysis of samples extracted from the building. Treated in 1975 following those specifications the building appears to be largely in a good state of preservation. Methods to remove efflorescence and consolidate friable stone is described with the example of the Sphinx.

REFERENCES

Postgate, J. R., *The Sulfate-Reducing Bacteria*, 2nd ed. Cambridge University Press: New York, 1984.

Selwitz, C., *Epoxy Resins in Stone Conservation*, The Getty Conservation Institute Research in Conservation, I. Averkieff (ed.), Los Angeles, 1992, Vol. 7.

CHAPTER 11

Geoarchaeology and the Age of the Sphinx

11.1 INTRODUCTION

Geoarchaeology is a discipline at the boundary between geology and archaeology. It is considered commonly as being the application of geological principles and techniques to the solution of archeological problems. Of late, however, it has developed to include the use of archaeological data to elucidate types and rates of recent geological processes. The question of the age of the Sphinx, however, has been highly puzzling. Archaeology has looked to geology for a tenable evidence, but the geological indicators of absolute age, such as charcoal and pottery, are absent from the Sphinx rocks or the surrounding materials. We believe, therefore, that the archaeological evidence available at this time is perhaps the best estimate.

The Great Sphinx of Giza is considered by Egyptologists to have been excavated by the Pharaoh Kephren in 26th century B.C. The egyptologic evidence for the age of the Sphinx is based on the view that Pharaoh Kephren created the Sphinx. This view, among other lines of evidence, derives from the setting of the Sphinx with respect to other neighboring pharaonic structures (Fig. 11.1). Kephren's Pyramid, the site of his burial, is located just west of the Sphinx. The valley temple, where he was anointed before burial, is situated immediately to the southeast, perhaps at the former bank of the Nile. The causeway, the route of Kephren's funerary procession skirts the south ditch of the Sphinx. Additionally, some Egyptologists believe that the face of the Sphinx resembles that of Kephren.

However, there has been a wave of popular opinion suggesting that the Sphinx is several thousand years older than thought and that it is the product and legacy of now vanished culture, a thesis now being supported by some scientists.

For example, geologist Schoch (1991) believes the Sphinx is older based upon the rounded profile of the strata of the Sphinx thorax and the deep channels present in the walls surrounding the Sphinx ditch. These geomorphologic features, he argues, are due to "precipitation-induced weathering," formed when the Sahara still experienced a humid climate at least 7,000 years ago. Later, Dobecki and Schoch (1992) presented geophysical evidence, based on seismic data on the weathering of the Sphinx

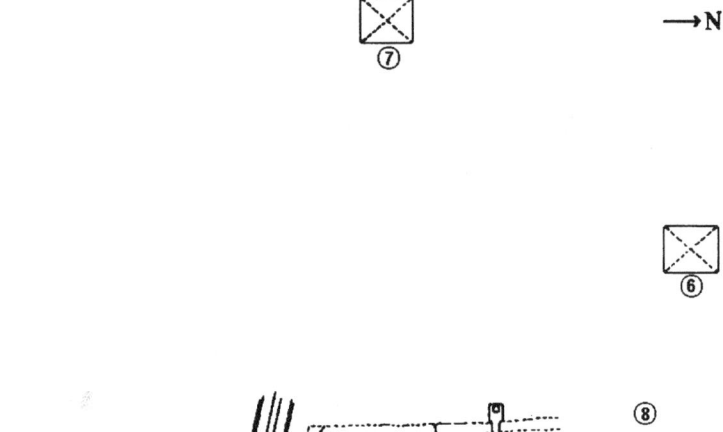

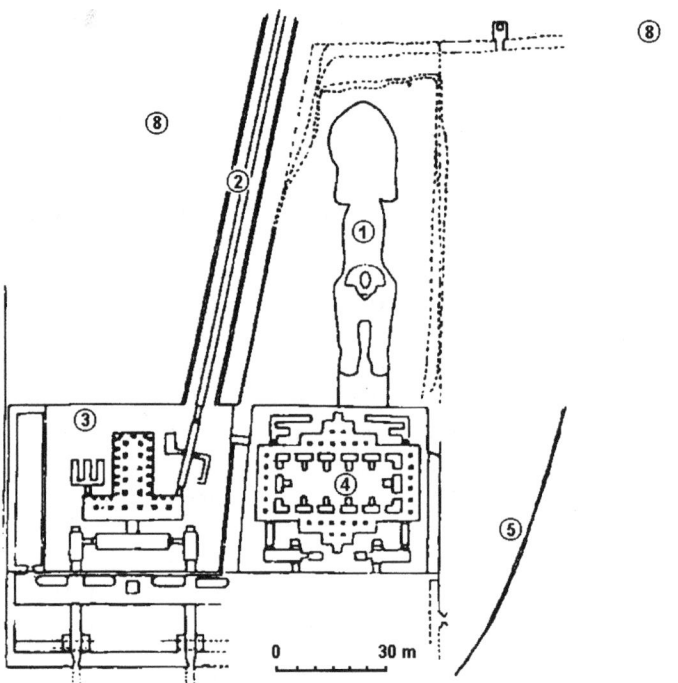

FIGURE 11.1 Setting of the Sphinx among other important pharaonic monuments. The distances and the symmetry of the pyramids relative to Sphinx are changed in order to show all features on one page. The numbers 1 to 8 in the figure represent (1) Sphinx, (2) causeway, (3) Valley Temple, (4) Sphinx Temple, (5) road to pyramids, (6) Khufu Pyramid, (7) Kephren Pyramid, (8) pharaonic necropolis. The pharaonic necropolis surrounds the Sphinx on all sides except east. *Source:* Gauri, K. L., Sinai, J. J., and Bandyopadhyay, J. K., Geologic weathering and its implication on the age of the Sphinx, *Geoarchaeology*, **10**(2): 119–133, 1995. Reprinted by permission of John Wiley & Sons, Inc. Copyright 1995 John Wiley & Sons, Inc.

subsurface, suggesting that at least the frontal portion of the Sphinx was excavated in prepharaonic times.

In this chapter we show how weathering in arid environment can produce the rounded profile and that the channels are actually the pre-Pliocene (older than 5 million years) karst features, formed by solution of limestone by the underground water, now exposed to surface due to pharaonic excavations at the Giza Plateau. This suggests that an absolute age cannot be assigned to the Sphinx based upon geomorphologic interpretations and seismic record. Thus one may have to rely on the previously presented circumstantial egyptologic evidence suggesting that the Sphinx is nearly 4,500 years old.

11.2 GEOPHYSICAL APPLICATIONS TO SPHINX ARCHAEOLOGY

Geophysical methods based on acoustic, electrical, dielectric, thermal, and magnetic properties of the rocks have been classically used in deep subsurface exploration for mineral deposits. In recent times, however, certain geophysical methods have been developed to explore the near-surface morphology. The applications of such methods have aided the science of archaeology by identification of archaeological sites and underground caverns.

Some such geophysical investigations dealing with the exploration of Sphinx subsurface include the Stanford Research Institute (SRI) use of electrical resistivity (Dolphin, 1981) and the Waseda University (Yoshimura et al., 1987, 1988) application of electromagnetic wave propagation, ground-penetrating radar, and gravimetric survey. The main purpose of these investigations was to locate subsurface cavities, a field of interest to Egyptologists. Lately, Dobecki and Schoch (1992), using seismic (compressional or acoustic) wave propagation velocity, classified the subsurface of Sphinx area and related it to the surface weathering and the age of the Sphinx.

Seismic refraction mapping depends upon the velocity of propagation of seismic waves in different media and upon an abrupt change of velocity, called seismic discontinuity, at an interface between two media of different physical properties. In general, the wave velocity can be expressed in terms of the density of the medium: the velocity is lower in a less dense medium and increases abruptly when the waves penetrate the underlying denser layers. A lower velocity horizon can, therefore, be indicative of weathered rock or a highly porous rock that may be saturated with water.

Dobecki and Schoch (1992) produced seismic tomography profiles (Fig. 11.2) showing that the depth of weathering along the length of the Sphinx, both in the north and south ditch, is somewhat similar. However the profile on the rear of the Sphinx reveals a nearly 1-m-thick low-velocity layer—interpreted as the weathered layer—while it is more than 3 m at places in the front. As the thickness of the zone of weathering is a function of time, they interpreted these data indicating that the frontal portion of the Sphinx must have been exposed to weathering for a longer period. This led to their conclusion that the front of the Sphinx must have been excavated earlier than the rear; the latter, which they do not dispute, as having been excavated during pharaonic times

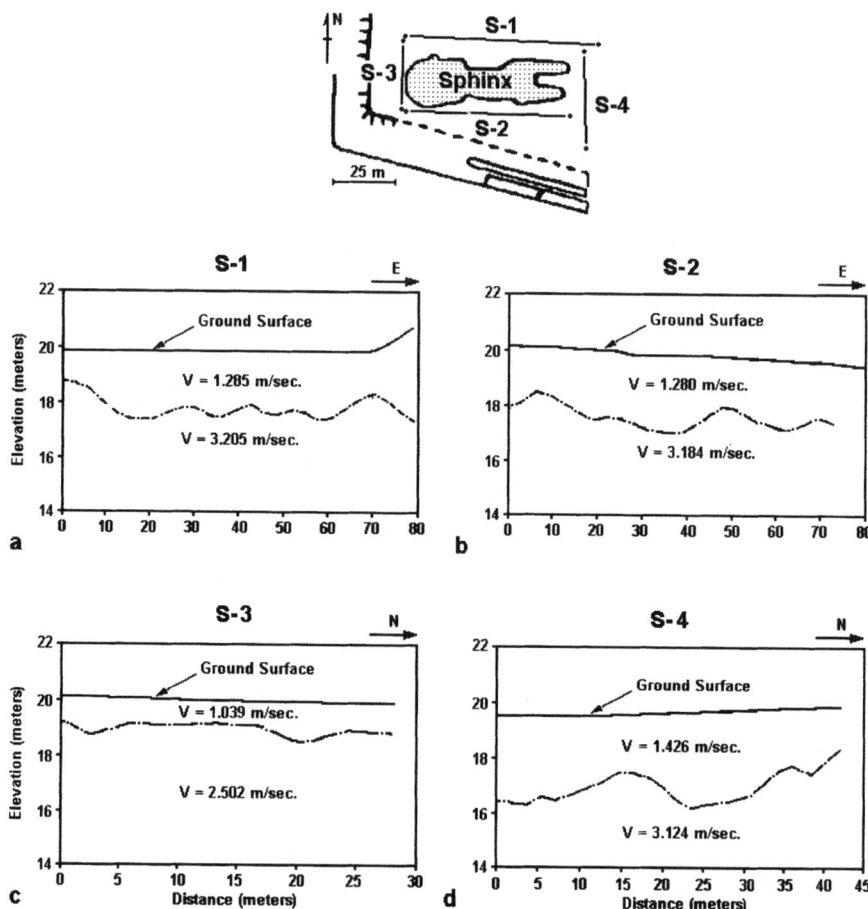

FIGURE 11.2 Seismic refraction profile of the subsurface around the Sphinx showing profiles along lines marked S-1 to S-4, where S-4 and S-2 are in the north and south of the Sphinx and S-4 and S-3 are in the east and the west respectively. The low-velocity zone beneath the ground surface is considered as the zone of weathering and the high-velocity zone beneath it is the unweathered country rock. *Source*: Dobecki, T. L. and Schoch, R. M., Seismic investigation in the vicinity of the Great Sphinx of Giza, Egypt, *Geoarchaeology* 7(6): 527–544, 1992.

11.3 WEATHERING AND THE AGE OF THE SPHINX

Schoch (1991) developed a theory on the age of the Sphinx based on magnitude and features of weathering and used the seismic refraction data given above in an attempt to corroborate that theory. According to his theory, the age of the Sphinx must be at least 7,000 years as compared with nearly 4,500 years considered by the Egyptologists. "The keystone of this theory is the nature and degree of chemical and mechanical weathering which the limestone walls of the sphinx enclosure, the body of the Great Sphinx, and the limestone cores of the Sphinx have experienced since becoming subaerially exposed."

Dobecki and Schoch (1992) interpreted the modern weathering profile of the Sphinx as a result of wet climate rather than the dry climate as it has prevailed since pharaonic times. Now, it is known that the Sahara experienced wet climate during what is called the Holocene pluvial. Holocene is the geologic time (Fig. 1.3) that followed the Pleistocene ice age and pluvial is a period of abundant rain associated with general cooling of the earth's climate due to the ice age. The Holocene pluvial ended nearly 7,000 years ago; thus, Dobecki and Schoch assert that the Sphinx must be at least 7,000 years old. The weathering during that period, "water-induced weathering" as they call it is, as they note, expressed by the following features observed at the Sphinx:

- Ravinelike structures present in the walls surrounding the Sphinx ditch
- Rolling vertical profile of the Sphinx
- Deep weathering of the Sphinx

We will discuss the origins of these features as we understand them in the following and demonstrate the error of these interpretations of the geomorphology of the area.

11.3.1 Ravinelike Structures

These structures exist along structural joints in the rock (Fig. 11.3). A careful observation of the rock surrounding joints shows colored bands that converge downward toward a joint (Fig. 11.4). These bands represent the receding underground water table as the Sphinx strata were gradually uplifted from the sea floor. In other words, the ravines had formed in the subsurface millions of years ago probably in the pre-Pliocene by the dissolution of limestone along joints. This phenomenon is, in fact, similar to that considered responsible for the formation of caves in general in a limestone terrain. No doubt the ravine walls have been widened somewhat since they were exposed during the carving of the Sphinx, but their origin dates back to geologic antiquity. This conclusion is further confirmed by a recent excavation (Fig. 11.5) near the Khafre Pyramid.

11.3.2 Rolling Vertical Profile of the Sphinx

Figure 11.6 shows what Schcoch and Dobecke appear to have meant by the rolling profile. We have shown in Chapter 8 that this profile has been created due to the differential weathering of the strata controlled by the makeup of the rock.

It is widely known that the rocks in the area surrounding the Sphinx were excavated for burial during pharaonic time, a time when the climate had already become dry. Doebecke and Schoch (1992) postulate that the exposed vertical rock cliffs of this area—called pharaonic necropolis—acquired angular weathering profiles made by the wind action. But, a careful tracing of these strata (Fig. 11.7) shows that they are

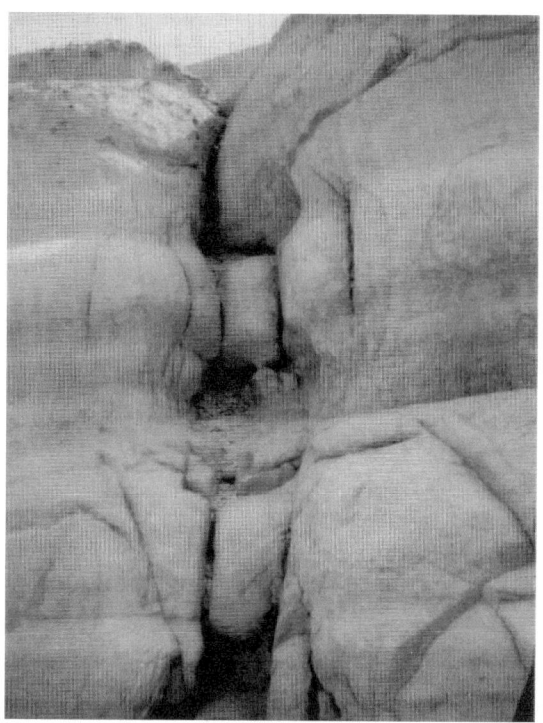

FIGURE 11.3 Channels and cavities in the Sphinx limestone. These dissolution features along structural joints were created by the underground water. They are now exposed at the surface due to the excavation of the Sphinx from the country rock.

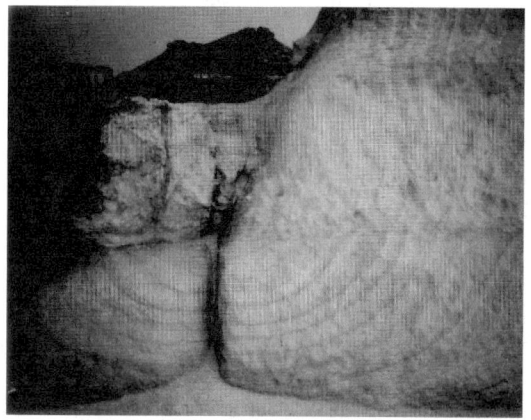

FIGURE 11.4 Color bands surrounding joints showing the downward convergence of the color bands towards joints, indicating the path of recession of underground water as the Sphinx strata were gradually uplifted above the sea floor in geologic time.

FIGURE **11.5** Underground cavities in limestone. These cavities, like those in Figure 11.4, had formed by the dissolution of rock by underground water and were exposed in this cut made in recent time. The site of this photograph is immediately north of the Khufu Pyramid shown in Figure 11.1.

FIGURE 11.6 Rolling vertical profile of the Sphinx. This profile resulted from the differential weathering due to the change in lithology of alternating starta. The harder layers made of relatively purer limestone are projecting outward due to less weathering than the softer layer, which have weathered more and are deeply recessed.

FIGURE 11.7 Rock layers at the Sphinx and their extension in pharaonic necropolis. Aerial view showing bed 4 of the Sphinx as the whitish surface in the right middle part of the photograph. Notice that this and other units of the Sphinx are buried under the sand towards the south (towards bottom of picture) and that younger layers of the Sphinx torso and head are deep underneath the layers revealed in the pharaonic necropolis towards the west. *Source*: Gauri, K. L., Sinai, J. J., and Bandyopadhayay, J. K., Geologic weathering and age of the Sphinx, *Geoarchaeology* **10**(2): 119–133, 1995. Reprinted by permission of John Wiley & Sons, Inc. Copyright 1995 John Wiley & Sons, Inc.

FIGURE 11.8 A rolling profile in the necropolis area. Notice the similarity of this profile with the Sphinx profile shown in Figure 11.6. *Source*: Gauri, K. L., Sinai, J. J., and Bandyopadhayay, J. K., Geologic weathering and age of the Sphinx, *Geoarchaeology* **10**(2): 119–133, 1995.

FIGURE 11.9 Rapid modern weathering in the Sphinx area showing deep weathering of a stone block that was exposed for less than one year at the floor of the Sphinx ditch. *Source*: Gauri, K. L., Sinai, J. J., and Bandyopadhayay, J. K., Geologic weathering and age of the Sphinx, *Geoarchaeology* **10**(2): 119–133, 1995.

FIGURE 11.10 Sphinx buried under the desert sand showing a lithographic reproduction from Description de L'Egypte. As shown in this view, most of the Sphinx had been buried for a longer period than it was exposed to the elements.

younger than those present at the Sphinx. Most of these strata have a somewhat different petrologic makeup. Nonetheless, one finds outcrops in the necropolis that show a profile identical to that of the Sphinx (Fig. 11.8).

Furthermore, the rate of weathering in desert area can be very fast depending upon the makeup of the strata. Figure 11.9 shows a block of limestone left in the Sphinx ditch in 1978, intended to be used next year in the restoration of the veneer. Within a year the stone surface had peeled off, obliterating even the tool marks. A conclusion that one can draw from this is that the rate of weathering of certain strata has been so fast in recent times that the presumed prephaoranic rounded profile would be superimposed by the subsequent angular profile.

11.3.3 Deep Weathering of the Sphinx

The Sphinx had been buried under the desert sand for most of the time of its existence, but not accurately determinable (Fig. 11.10). Schoch and co-workers were awed at the intense weathering that the Sphinx had undergone through its life. They surmised that the Sphinx must have been exposed for a time longer than thought in order to have suffered that much degradation. Because the exact time for which the Sphinx was exposed to the elements and the rate of weathering are unknown, correlation of the degree of weathering of the Sphinx with its absolute age would be, at best, fortuitous.

SUMMARY

Monuments on the Giza Plateau are considered to be of pharaonic origin dating to early dynastic period that lasted between 2,920 and 2,575 B.C. However, new estimates extend the age of the Sphinx backward to Holocene Pluvial, which existed before 7,000 B.C. Unfortunately, common indicators of absolute age of archeological objects, such as pottery and charcoal, have not been found in association with the Sphinx. Therefore, any estimate of age has to be based on indirect evidence.

The evidence used by the authors of the new theory is based upon interpretation of geomorhological features and seismic data. We show in this chapter that deep channels in the walls of the Sphinx ditch that they consider as having been formed by running water are actually caves formed by the underground water in geologic antiquity. We also show that the rounded vertical profile of the Sphinx, which they consider as having been carved in the humid pre-Sahara, is being created by the present-day weathering. We show further vertical profiles of weathered rock, similar to that of the Sphinx, in the Giza necropolis, which is accepted even by the new authors as being pharaonic based on seismic evidence. Thus, they misinterpreted the geological features that are the basis of their theory.

Therefore, the Sphinx should continue to be regarded as of pharaonic age—excavated in 2,600 B.C.—as considered by Egyptologists until further evidence is forthcoming. At The First International Symposium on the Great Sphinx, 1992, sponsored by the Egyptian Antiquities Organization, Egyptologists wondered whether the Sphinx had been carved during the reign of Pharaoh Kephren, as is commonly believed, or during the reign of his father, the Pharaoh Khufu, whose pyramid stands nearby!

REFERENCES

Dobecki, T. L, and Schoch, R. M., Seismic investigations in the vicinity of the Great Sphinx of Giza, Egypt, *Geoarchaeology*, **7**: 527–544, 1992.

Dolphin, L., Application of novel remote-sensing techniques to advance the state of archaeology in Egypt, *Report of SRI Project No. 5669B*, Stanford Research Institute, Menlo Park, CA, 1981.

Schoch, R. M., Redating the Great Sphinx of Giza, Egypt, Abstracts with Programs, GSA (Geological Society of America) Annual Meeting, San Diego, 1991, A 253.

Yoshimura S., Tonouchi, S., Nakagawa, T., and Seki, K., Non-destructive pyramid investigation (1)—by electromagnetic wave method, *Studies in Egyptian Culture No. 6*, Waseda University, Tokyo, 1987.

Yoshimura S., Nakagawa, T., and Tonouchi, S., Non-destructive pyramid investigation (2), *Studies in Egyptian Culture No. 8*, Waseda University, Tokyo, 1988.

Artificial Neural Networks

A.1 INTRODUCTION

Artificial neural networks are computational tools the algorithms (rules) of which are based on the organization of the biological neural system consisting of billions of neurons. Each neuron may simply be considered as information storage and processing unit. The interactions between neurons are provided by *synapses*, which are gaps between neurons. In a *synapse*, the electrical signal generated in a neuron is converted to a chemical signal by releasing what are called *neurotransmitters*, which then convert the chemical back into an electrical signal when passing information onto the next neuron or neuron groups. The neurons are arranged in cascading hierarchial system. The synapses are either *excitatory* by which the signal is amplified or *inhibitory*, which reduce the signal. These fundamental aspects of the biological neural system are simulated in the artificial neural networks.

A.2 ARTIFICIAL NEURAL NETWORKS

In an artificial neural network (ANN), the input and the associated output data are fed. ANN may be considered as a black box containing one or more computational logic units (CLU) connected in tandem. The first and last layers in this network are input and output layers, respectively, with hidden layers present in between. While the input and target values are obtained from observation, the values in the hidden layers are found by training with the help of an external teacher that optimizes the values through error signals and bias corrections. When the training is complete, the converged network has the ability to recognize patterns intrinsic to the training data.

In the human neural system, the degree of response to a stimulus (an input) is believed to be a function of the connecting link, the synaptic strength, which is represented in ANN by weights. The weight is positive if the associate synapse is excitatory; it is negative if the synapse is inhibitory. To train ANNs, then, the inputs are multiplied by weights of random values, and in a multilayered network—a parallel architecture of several computational logic units in a stack and in tandem—weights are found that match the input with the target values using an error backpropagation algorithm. We will describe mathematically the methodology involved in a multilayer neural network.

The input (I), considered as a $n \times p$ matrix, where n is the number of inputs and p is the number of data sets present, is multiplied by randomly selected weights $W_{i,j}$. Conventionally, the first subscript i refers to the receiving neuron and second subscript j (here $j = 1, n$) denotes the input neuron. Thus the total inputs to the ith CLU in a hidden layer is given by

$$X_{i,p} = \sum_{j=1}^{n} W_{i,j} I_{j,p} + u_i \tag{A.1}$$

where u_i is the threshold or bias. We will explain later why bias is needed. The resulting number for the pth pattern in the ith CLU is processed further to produce an output as follows:

$$Y_{i,p} = f(X_{i,p}) \tag{A.2}$$

where the function f is any suitable functional form, known as the *activation function*. The activation function may be linear or nonlinear in nature. The most commonly used activation function is the sigmoid function, given as

$$f(X_{i,p}) = \frac{1}{1 + \exp(-aX_{i,p})} \tag{A.3}$$

where a is the slope parameter of the sigmoid function. Normally, the chosen value of slope parameter $a = 1$, but the assigned value may be in any range. If the value of a is infinity, the sigmoid function becomes a threshold function, yielding a value of either 0 or 1, like binary switch. For other values a the output becomes 0 to 1 in a continuous manner.

Using Eq. (A.3), Eq. (A.2) becomes

$$Y_{i,p} = \frac{1}{1 + \exp(-X_{i,p})} \tag{A.4}$$

The process is repeated for all the layers of the network. In the architecture used here we have one hidden layer in which the values obtained from Eq. (A.4) form the inputs to the output layer. Thus the total input to a particular neuron in the output layer will be given by

$$Z_{k,p} = \sum_{i=1}^{m} W_{2_{k,i}} Y_{i,p} + u_{2_k} \tag{A.5}$$

where W_2, and u_2 are the weight matrix and bias vector for the output layer. If we denote the output of the kth neuron in the output layer as $O_{k,p}$, then

$$O_{k,p} = \frac{1}{1 + \exp(-Z_{k,p})} \tag{A.6}$$

The computed output from the kth neuron, $O_{k,p}$, is now compared with the target value. The error e, defined as the difference between the computed output ($O_{k,p}$) from the network and the desired output ($T_{k,p}$) for a particular pattern p at the final layer, is calculated as

$$e_{k,p} = (T_{k,p} - O_{k,p}) \tag{A.7}$$

The error signal defined in Eq. (A.7) then becomes the basis for updating the weights. Customarily, the instantaneous value of squared error for the kth neuron is defined as $\frac{1}{2} e_{k,p}^2$, and the instantaneous error (E_p) for the pattern p is the sum of instantaneous squared error obtained by summing squared errors over all neurons in the output layer, given by the following expression

$$E_p = \frac{1}{2} \sum_k e_{k,p}^2 \tag{A.8}$$

The objective is to adjust the free parameters such as synaptic weights and bias vector, in the network architecture to minimize the error E_p (the objective function). To get an estimate of the learning performance of the network, the total error is computed as follows:

$$E_{\text{overall}} = \sum_p E_p = \sum_p \sum_k \frac{1}{2} e_{k,p}^2 \tag{A.9}$$

There are several methods for minimizing the errors. Here we will apply the steepest-descent algorithm for error minimization, and error back propagation (EBP) for weight corrections.

The weight adjustment between the originating neuron A and the destination neuron B is performed as

$$\Delta W_{BA}^{N+1} = \eta \left(\frac{\partial E_p}{\partial W_{BA}} \right) + \alpha \Delta W_{BA}^{N} \tag{A.10}$$

In order to simplify the mathematical rigor [for details of the derivation, see Haykin (1994)], we will explain the terms on the right-hand side of Eq. (A.10). The factor η ($0 < \eta < 1$) is termed as learning parameter, and the term $\partial E_p / \partial W_{BA}$ can be split as follows:

$$\frac{\partial E_p}{\partial W_{BA}} = \delta_B(N+1) \times V_A$$

$$\underbrace{\hphantom{\delta_B(N+1)}}_{\text{local gradient}} \quad \text{input signal to neuron B} \qquad \text{(A.11)}$$

The local gradient term will depend on whether the neuron B is in the output layer or in the hidden layer. If B is in the output layer, the local gradient of the kth neuron δ_k can be written as

$$\delta_k(N+1) = e_{k,p}(N)f'(Z_{k,p}) \qquad \text{(A.12)}$$

where, $f'(x)$ denotes the first derivative with respect to x, and

$$V_A = Z_{k,p} \qquad \text{(A.13)}$$

where N is the iteration number.

If the neuron B is in the hidden layer then the local gradient for Hth neuron δ_H is calculated as

$$\delta_H(N+1) = f'(Y_{i,p}) \sum_k \delta_k(N+1)W_{k,H} \qquad \text{(A.14)}$$

and

$$V_A = Y_{i,p} \qquad \text{(A.15)}$$

The elements in the right-hand side of Eq. (A.12) can be given as

$$f(Z_{k,p}) = \frac{1}{1 + \exp(-Z_{k,p})}$$

$$\Rightarrow f'(Z_{k,p}) = \frac{\exp(-Z_{k,p})}{[1 + \exp(-Z_{k,p})]^2} \qquad \text{(A.16a)}$$

Using Eq. (A.6) we can write

$$1 - O_{k,p} = \frac{\exp(-Z_{k,p})}{1 + \exp(-Z_{k,p})} \qquad \text{(A.16b)}$$

Combining Eq. (A.16a) with Eq. (A.16b), and Eq. (A.6)

$$f'(Z_{k,p}) = O_{k,p}(1 - O_{k,p}) \tag{A.17}$$

Note Eq. (A.17) has been expressed in terms of output signal. Using the error term given by Eq. (A.7), the local gradient becomes

$$\delta_0(N + 1) = (T_{k,p} - O_{k,p})O_{k,p}(1 - O_{k,p}) \tag{A.18}$$

and the final form to update the output layer weight is

$$\eta \frac{\partial E_p}{\partial W_{BA}}(N + 1) = \eta(T_{k,p} - O_{k,p})O_{k,p}(1 - O_{k,p})Z_{k,p} \tag{A.19}$$

In the hidden layer, weight is updated as shown in Eq. (A.14). Following the derivation of Eq. (A.16), we can write

$$f(X_{i,p}) = \frac{1}{1 + \exp(-X_{i,p})}$$

$$\Rightarrow f'(X_{i,p}) = \frac{\exp(-X_{i,p})}{[1 + \exp(-X_{i,p})]^2} \tag{A.20}$$

Combining Eq. (A.5) with Eq. (A.19) we get

$$f'(X_{i,p}) = Y_{i,p}(1 - Y_{i,p}) \tag{A.21}$$

So the local gradient for hidden layer becomes

$$\delta_H(N + 1) = Y_{i,p}(1 - Y_{i,p}) \sum_k \delta_k(N + 1)W_{k,H} \tag{A.22}$$

and the final form for updating the hidden-layer weight can be written as

$$\frac{\partial E_p}{\partial W_{H,A}}(N + 1) = \left(Y_{i,p}(1 - Y_{i,p}) \sum_k \delta_k(N + 1)W_{k,H} \right)X_{H,p} \tag{A.23}$$

Returning to Eq. (A.10), the second term on the right-hand side is known as the *momentum* term and is used to accelerate the learning rate by adding a fraction of immediate past weight. The coefficient α $(0 < \alpha < 1)$ is known as momentum constant parameter.

The bias term $(u, u_2,$ etc.) is added to the sum total before squashing. This is due to the fact that sometimes the output after squashing may be in the intermediate range,

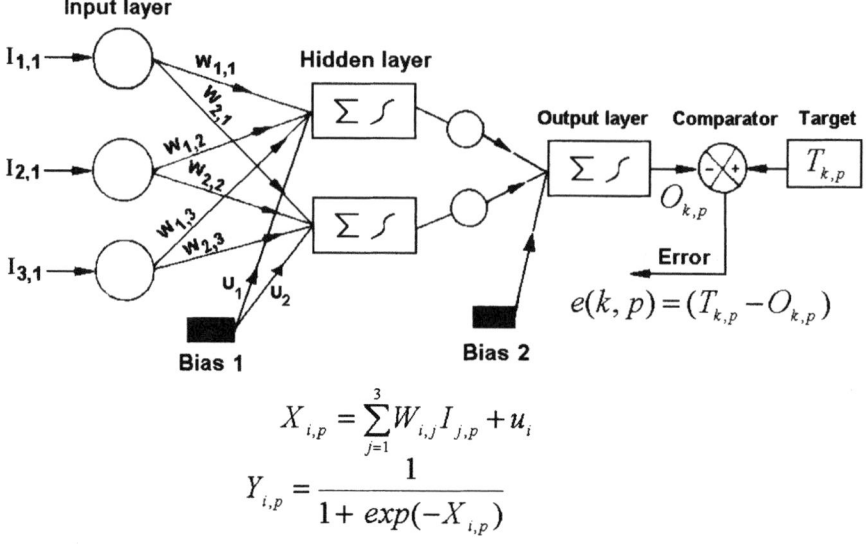

$$X_{i,p} = \sum_{j=1}^{3} W_{i,j} I_{j,p} + u_i$$

$$Y_{i,p} = \frac{1}{1 + exp(-X_{i,p})}$$

FIGURE A.1 Architecture of an artificial neural network.

neither close to 0 nor 1. These are called noncommitted neurons. Adding a bias will make these noncommitted neuron become a committed neuron by making its value either closer to 0 or 1. Following the method describe above, the weights are adjusted until the error value is less than some predefined value.

Figure A.1 gives the architecture of ANN and gives salient equations at sites where related processing is performed. It should be noted that too many neurons in the hidden layer may cause the architecture to memorize the pattern instead of generalization.

REFERENCE

Haykin, S., *Neural Networks—A Comprehensive Foundation*, New York: Macmillan College Publishing, 1994.

Water Vapor in the Earth's Atmosphere

The volume of water vapor in the Earth's atmosphere varies from negligible to 4%. Yet, water vapors control nearly the entire process of stone weathering. For example, adsorption of water by salts and clay minerals, as discussed in Chapters 4 and 9, is the main cause of the mechanical disintegration, and, as further shown in Chapter 9, condensed moisture is the only source of water responsible for weathering in a desert climate. Furthermore, we saw in Chapters 5 and 6 that the reaction rates of carbonate minerals by dry deposition are directly proportional to the prevailing relative humidity and dew, and that the oxidation of iron bars in masonry structures is also supported by water. In this appendix, we explain some of the terms used in hygrometry. We also present a psychrometric chart (Fig. B.1) whereby moisture-related properties, such as relative humidity and dew point often mentioned in the previous chapters, can be determined.

Relative humidity (Rh) is the ratio of the amount of water vapor present in a sample to the water vapor needed to saturate that sample. Rh is often expressed as a percentage. For example, saturated air at 20°C (68°F) has a water vapor pressure[*] of 2.338 kPa (kilopascal). Air, with vapor pressure of 1.169 kPa, has at the same temperature the relative humidity of 50%.

Dew point is the temperature at which the relative humidity reaches 100% and the condensation begins.

The dew point is determined by an instrument called a *psychrometer* (it relates to cooling caused by the evaporation of water). A psychrometer is a hygrometer consisting essentially of two thermometers: One records the air temperature (dry-bulb thermometer), and the other has its bulb covered with muslin that wicks water and cools the surface of the bulb (wet-bulb thermometer). Psychrometric charts (Fig. B.1) can be used to determine dew point and relative humidity for desired dry-bulb temperature and wet-bulb temperature. Our ftp site contains a program whereby dew point can be calculated if air temperature and relative humidity are known.

---

[*]The volume of water vapor is often expressed as vapor pressure. The term *vapor pressure* refers to the partial pressure exerted by water vapor in the atmosphere. The pressure exerted by the standard air, that is air with all its gases, is one atmosphere, which is equivalent to 101.325 kPa [= 1.013 bars (1 bar is the pressure of 1 kg/cm^2) = 1013.25 millibars = 760 mm = 29.9 inches of mercury].

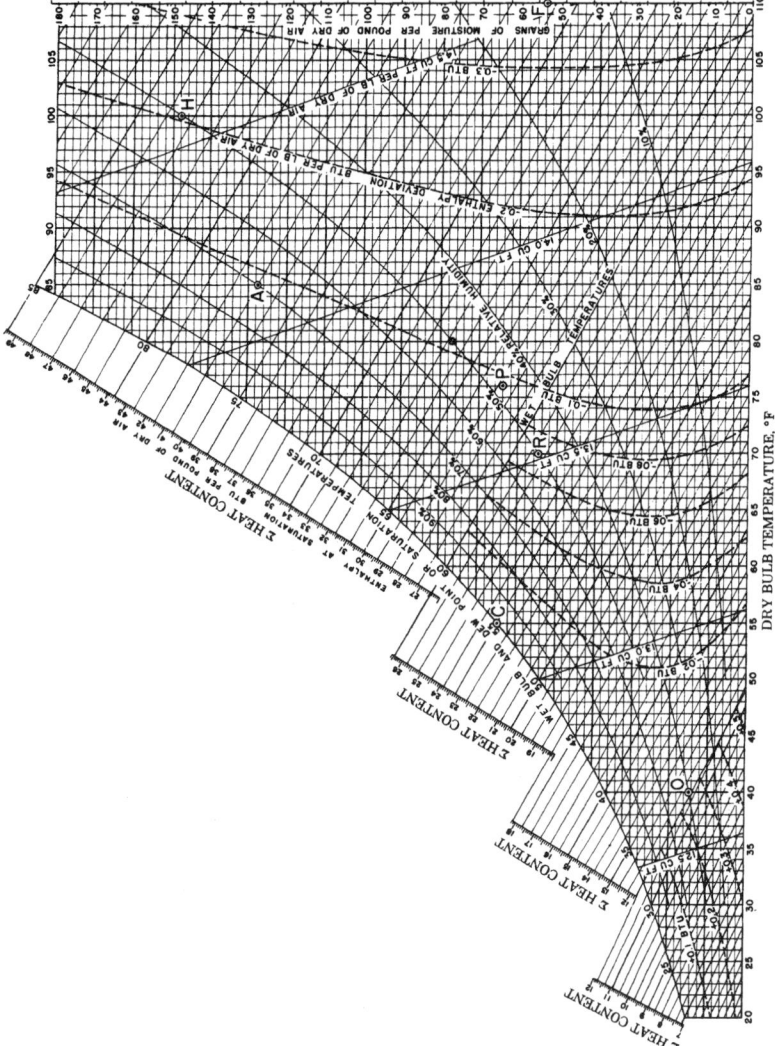

FIGURE B.1 Psychrometric chart. Wet bulb temperature and air temperature are shown along with relative humidity curves. To determine relative humidity, for example, if the dry-bulb reading is 80°F and the wet-bulb reading 70°F, then the Rh is 60%. Further, if the relative humidity and room temperatures are known as being 60% and 80°F, then the saturation temperature (dew point) is 65°F.

Web Component of the Book

The Web component of the book contains computer programs by which properties, decay rates, and durability factors of limestone can be calculated. The programs allow calculation of the following:

Durability factors by regression analysis
Durability factors and chemical rates by artificial neural networks
Rate-related constants
Porosity by water absorption
Surface reduction by rain
X-ray diffraction 2θ and d-spacing
Relative humidity and dew point

These computer programs can be accessed via the Internet at the following address: ftp://ftp.wiley.com/public/sci_tech_med/limestone.

GLOSSARY

Acid rain: Rain with pH below 5.6 due mainly to the solution of atmospheric SO_2 and NO_2.

Acidolysis: Solution of carbonate minerals in acids produced by microbes.

Acropolis of Athens: Takes its name from Greek words *akros* (upper) and *polis* (city). Built nearly 2,500 years ago, the temples on Acropolis have suffered weathering in the present century due to anthropogenic pollutants and human endeavors to protect the stone from future decay.

Aerobic organism: Organisms that can exist in an oxygenated environment only.

Algal mat: Layered growth of algae on intertidal and subtidal zones, which may cause carbonate sedimentation.

Allochem: Coarse particles, larger than 10 µm, in carbonate sediment or rock. Allochem consists of shell fragments, ooids or ooliths, fecal pellets, and intraclasts.

Anaerobic organism: Organisms that can survive only in an oxygen-devoid atmosphere.

Ångstrom: Unit of length measurement equal to 10^{-10} m.

Anion: Negatively charged ion.

Anticline: Fold in a rock in which the limbs slope away from the axis.

Aragonite: Mineral polymorph of $CaCO_3$.

Artificial neural network: Mostly known by the abbreviation ANN, it is a mathematical model that can learn known patterns between input and output and recognize these if present in unknown data. It is designed to simulate biological neural system, hence the name artificial neural network.

Atmosphere: Atmosphere is the gaseous envelope that surrounds a planet. Earth's atmosphere is a mixture of gases, predominantly nitrogen (78%), oxygen (21%), and water (0.5–4%). However gases with much lower concentration namely CO_2 (360 ppm), SO_2 and NO_2 (ppb to ppm level) are most effective in material's decay.

Atmosphere: Unit of pressure equal to 101,325 newtons per square meter or about 14.7 pounds per square inch.

Attitude: Term that comprises strike and dip. A rock layer inclined to the earth's surface has a *strike* direction that is horizontal and a *dip* direction in which the rock slopes away from the horizontal. Thus, attitude refers to structurally deformed rocks because nondeformed rocks are often horizontal.

Autotroph: Organisms that manufacture food independently by photosynthesis (photoautotrophs) or by "breathing" complex ions, such as sulfate and nitrate (chemoautotrophs).

Basin: Site of accumulation of a large thickness of sediment.

Bedding or stratification: Characteristic of sedimentary rocks in which layers of different grain size or composition are separated by planar surfaces, called bedding planes.

Bicarbonate ion: Anion group HCO_3^-.

Biochemical precipitate: Sediment, especially of limestone, formed from elements extracted from seawater by living organisms.

Biodeterioration: Stone decay, particularly due to microflora (algae, fungi, lichens, mosses, and bacteria).

Biodeterogens: Organisms responsible for biodeterioration.

Biofilm: Microbial encrustation upon a stone surface.

Bioherm: Mass of carbonate reef built on the sea floor exclusively by many organisms, including corals, foraminifers, echinoderms, and mollusks. Bioherms lack bedding, which is common in most limestone.

Boundary layer: A layer of air surrounding any solid surface held motionless by friction.

Bragg equation: Fundamental equation for X-ray diffraction: $n\lambda = 2d \sin \theta$, where λ is the wavelength, d is the spacing between atomic planes, and θ is the reflection angle.

Brittle material: Solid that when subjected to compressional stress shortens at first, called elastic strain, but breaks instantaneously if the stress exceeds the limit of elastic strain.

Bulk density: See density.

Calcite: Polymorph of $CaCO_3$.

Calcite, high- and low-magnesian: Calcite with a magnesium carbonate ($MgCO_3$) content of more and less than 4%, respectively.

California Building: Building in San Diego presented in this book as a case history of conservation treatment.

Capillary: Narrow pore.

Carbonate ion: Anion group CO_3^{2-}.

Carbonate platform: Submarine or intertidal shelf the elevation of which is maintained by active shallow-water carbonate deposition.

Carbonate rock: Rock composed of carbonate minerals. Common carbonate rocks are limestone, dolostone, and marble. Limestone is the parent carbonate rock; dolostone forms by its chemical replacement and marble by its metamorphism.

Carbonate sediment: Sediment deposited largely by biochemical processes in a marine environment; precursors to all carbonate rocks.

Carbonic acid: Weak acid H_2CO_3, formed by the dissolution of CO_2 in water.

Caryatids: Maiden figures in architecture used in place of columns originally at the temple of Erechtheum at the Acropolis.

Cation: Any ion with a positive electric charge.

Chelation: Removal of cations from minerals by organic acids produced by microbes, causing breakdown of the mineral.

Chemical differentiation: Formation of different minerals from magma as it cools.

Chemical sediment: One that is formed at or near its place of deposition by chemical precipitation, usually from seawater. Gypsum ($CaSO_4 \cdot 2H_2O$) and halite ($NaCl$) are common chemical sediments.

Chemical weathering: All chemical reactions that act on rock exposed to water and atmosphere and so change its minerals.

Clastic particles: Mineral particles (clasts) that were mechanically transported. In carbonate rocks they consist of mostly clay minerals and silt.

Clay minerals: Hydrous aluminosilicates formed by hydration of primary silicate minerals; also, any mineral fragment smaller than $1/256$ mm (4 μm).

Clean rain: Rain with pH 5.6 or higher, indicating the absence of pollutant gases in the atmosphere. The pH 5.6 is obtained by the dissolution of atmospheric CO_2.

Cleaning: In stone conservation, cleaning refers to treatment of crust and removal of efflorescence, stains, and street dust.

Compaction: Decrease in volume and porosity of sediment caused by deep burial.

Composite material: Construction material for patching, repair, or replacement of worn-out portions of building; often formulated by the combination of polymeric adhesive and crushed rock to simulate the natural rock.

Compressive strength: Force at which a solid will break when under compressive stress.

Conservation: In reference to architectural objects conservation comprises the diagnosis of the problem, treatment, preventive measures, and restoration.

Consolidation: Hardening of weathered rock that has partially lost cement, often by impregnation with organic polymers.

Contact angle: Angle that the surface of a drop of liquid makes with the substrate. The angle depends upon the surface tension of the liquid and the electrochemical nature of the capillary wall. Carbonate rocks have contact angle less than 90° with water.

Continental shelf: Shallow portion of the sea floor adjacent to land.

Cross-bedding: Inclined layers between parallel bedding planes of depositional origin in a sedimentary rock. Formed by wave action during the formation of carbonate sediment.

Crust: Appendage upon a stone surface made of the combination of the reaction product of chemically active atmospheric gases with the stone substrate and the atmospheric dust that becomes entrapped as the reaction takes place.

Crystallization pressure: Pressure generated in pores due to salt crystallization during drying of wet rock.

d **spacing:** In X-ray diffraction, the space between adjacent atomic planes that produce diffraction.

Deliquescent: Salts are deliquescent when they absorb moisture from air and turn it to liquid water at a relative humidity below the dew point.

Density: Also called bulk density, it is the weight per unit volume (g/cm^3) of the sample. Skeletal density is the ratio of weight to its volume without pores.

Deposition velocity: Average rate of deposition of a gas towards an absorbing surface. It is a bulk coefficient combining the effects of mass transfer and surface effect; often expressed in cm/h.

Desulphovibrio desulphuricans: Bacteria that are believed to have converted anhydrite (CaSO$_4$) to calcite on a geological scale and used in cleaning gypsum crust from marble.

Detrital sediment: Sediment consisting of clastic particles.

Diagenesis: Processes that consolidate soft sediment to solid rock by *compaction* by pressure of overlying sediment, *cementation* by chemically precipitated primary calcite, *recrystallization* of secondary minerals, and *replacement* of some calcium in calcite by magnesium.

Diffusion, external diffusion or mass transfer: Transfer of reactant gas through a boundary layer to the sample surface, thus a rate-controlling parameter; expressed in cm/h.

Diffusion, internal: Transport of calcium ions from a stone's interior to the sample surface, a rate-controlling parameter; expressed in cm^2/h.

Dip: Angle that an inclined surface, such as a bedding plane or a joint, makes with the earth's surface.

Dolomite: Mineral CaMg(CO$_3$)$_2$, formed by replacement of some Ca in calcite by Mg in diagenetic processes.

Dolostone: Carbonate rock containing more than 50% dolomite. Dolostone is sometimes called dolomite.

Dry deposition: Deposition of environmental gases and particles upon stone surfaces, which are protected from rain.

Ductile rock: Rock that will flow but not break in increasing *stress* after the limit of *elastic strain* is reached.

Durability: Often considered as the ability of stone to withstand mechanical stress produced in the pores by, as an example, crystallization of salt.

Durability factor: Number, sometimes in the scale of 1 to 10, expressing the relative durability of a stratum.

Duricrust: Commonly referred to soil, it is the crust formed in arid climates by the precipitation of salts as the groundwater evaporates on the soil surface: a phenomenon termed *case hardening*. In stone conservation, it refers to the crust of minerals that are slightly soluble in water, such as gypsum, which remain preserved in dry climate and thus protect the underlying stone from further decay. In humid climates, recrystallized calcite from the parent limestone can also form such a crust.

Echinoderm: Spiny-skinned invertebrates, characterized by a radial and five-rayed symmetry of skeletal plates made of calcium carbonate. Their parts are found ubiquitously in limestone.

Effervescence: Bubbling caused by the release of carbon dioxide when an acid is dropped on a carbonate rock.

Efflorescence: Powdery salt upon stone or brick transported from within the stone. Efflorescence is the major cause of mechanical weathering of stone.

Elastic modulus: See modulus of elasticity.

Elastic strain: Maximum stress that can be applied to a body without resulting in permanent strain.

Entrapment hysteresis: In mercury porosimetry, a smaller volume of mercury is extruded than intruded in the initial intrusion, signifying entrapment in large pores, thus giving the volume of large pores in a rock.

Environment of deposition: Refers to the hydraulic condition in which carbonate sediment is deposited. A high-energy environment is one in which allochemical particles can be deposited, and a low-energy environment is one from which they are winnowed away.

Epeirogeny or epeirogenic movement: Large-scale vertical movements that lift the crust, sometimes producing fractures and joints in rocks.

Epoch: Subdivision of a geologic period.

Epsomite: $MgSO_4 \cdot 7H_2O$ forms by the reaction of sulfur dioxide with dolomite but is not preserved in crust on dolostone due to its deliquescence.

Era: Time period including several periods, but smaller than an eon. Commonly recognized eras are Precambrian, Paleozoic, Mesozoic, and Cenozoic.

Erosion: Processes by which loosened rock is moved away from a site.

Etching: Differential corrosion of a surface produced by an acid. Etch patterns can be used to differentiate different patterns of textures and minerals in carbonate rocks. For example micrite can be distinguished from allochem and sparry calcite by its dullness and most dolomites can be recognized from limestone by rhombohedral etch patterns.

Evaporite: Chemical sedimentary rock consisting of minerals precipitated by evaporating waters, especially halite and gypsum.

Exfoliation: Process in which crust is fractured and detached from the core.

Exfoliation joints: Joints in a rock parallel to the earth's surface formed by the removal of overburden, producing a release of confining pressure, especially when quarrying is done at a fast pace and in nature when deep-seated rocks are exposed by erosion.

Ferrous iron: Divalent iron is ferrous iron and has a brown color compared with trivalent ferric iron, which has a red color. The oxidation state of the iron present in a rock determines its color.

Fold: Folds are structural features of rock produced by compression in the earth's crust.

Foraminifera: Order of oceanic protozoa, most of which have shells composed of calcite.

Foraminiferal ooze: Calcareous sediment composed of the shells of dead Foraminifera.

Fossil fuel: Combustible geologic deposits of carbon, including coal, oil, natural gas, oil shales, and tar sands.

Fractal dimension: See power law.

Fractal structure: Structures that fall between the Euclidean linear and planar or planar and three-dimensional forms. Pores in rocks have fractal structure and each pore type can be characterized by a fractal dimension between 2 and 3 depending upon the geometry of pores.

Frost action: Rupture of rock by freeze–thaw activity.

Geoarchaeology: Discipline that applies geological principles and techniques to the solution of archaeological problems.

Geologic time scale: Division of geologic time in eras, periods, etc., and their approximate absolute age (see Fig. 1.3).

Geomorphology: Science of surface landforms and their interpretation on the basis of geology and climate or underground water.

Geophysics: Discipline that applies physics to the solution of geological problems, including exploration of the subsurface by seismic refraction.

Geyser: Hot spring that throws hot water and steam into the air. The heat is thought to result from the contact of groundwater with magma bodies.

Glyceration: Treatment of a clay mineral with glycerol solution whereby the spacing d can be enlarged. Used as a means to identify clay minerals.

Glycolation: Exposing a clay mineral mount to ethylene glycol vapor whereby the spacing d can be enlarged. Used as a means to identify clay minerals.

Hard limestone: Competent limestone in which clay minerals are nearly absent. When structurally deformed, such limestone acquires well-defined fractures or joints compared with incompetent limestone in which innumerable, minute fractures come into being.

Hardness: Relative scratchibility of minerals; a mineral that can scratch the other is harder. Mohs' scale of hardness (1–10) is such an empirical scale.

Heterotroph: Organisms that get their energy by eating autotrophs.

Holocene pluvial: Period of wet climate following the last ice age (Pleistocene), which culminated nearly 10,000 years ago.

Hot spring: Spring whose waters are above both human body and soil temperatures as a result of contact with magma at depth.

Hydration: Chemical reaction, usually in weathering, which adds water or OH^- to a mineral structure. Clay minerals form mainly by hydration of primary silicate minerals.

Hydration pressure: Pressure generated when an anhydrous salt becomes hydrated due to absorption of water vapor or while crystallizing from solution.

Hydration water: Highly ordered water that remains unfrozen in fine capillaries even at extremely low subfreezing temperature.

Hydrophilic: Materials with which water makes a small contact angle so that the pore wall readily attracts water and becomes wet. Carbonate rocks are hydrophilic.

Hydrophobic: Materials that do not attract water and remain dry when in contact with water due to their large contact angle with the substrate.

Hysteresis: In mercury porosimetry, (1) the difference in the volume of initial intrusion and extrusion of mercury and (2) extrusion and reintrusion at a given pressure.

Hysteresis loop: In mercury porosimetry the loop between cumulative plots of extrusion and reintrusion, formed due to the difference in the volume of extruded mercury and reintruded mercury at a given pressure.

Indiana limestone: Also called Bedford limestone of Carboniferous age, it is the most common building stone in the United States and is known for its high durability.

Initial intrusion, reintrusion, and extrusion: In   mercury   porosimetry   the sequential process of intruding mercury into stone pores with increasing pressure, extruding mercury by lowering the pressure, and then reintruding by an increase of pressure.

Ink-bottle pores: Pores in limestone characterized by large bellies, called large pores, connected to each other through narrow necks, which are called small pores.

Ion: Atom or group of atoms that has gained or lost electrons and so has a net electric charge.

Ion chromatography: An instrumental method for measuring ions such as sulfate and nitrate quantitatively.

Isomorphous replacement: Replacement of some aluminum and silicon in clay by magnesium and other ions producing a large variety of clay minerals.

Joint: Planar fracture in a rock across which there is no relative displacement of the two sides.

Kaolinite: Clay mineral.

Karst topography: Landforms due to solution of carbonate rocks, characterized by caves underground and sinkholes on earth's surface.

Kinetics of reaction: Study of reaction rates based on fundamental controls, such as deposition velocity of the reactive gas on the reactant and temperature.

Laurel dolomite: Geologic formation of Silurian age widely outcropping in the Louisville, Kentucky, area.

Lichen: Symbiotic association of algae and fungi.

Limestone: Sedimentary rock composed principally of calcium carbonate ($CaCO_3$), usually as the mineral calcite. Carbonate sediments become carbonate rocks by processes referred to as diagenesis.

Limonite: Iron hydroxide ($FeOH_3$), in weathering, produced by electrochemical reaction. Such oxidation of iron anchors used in masonry is a major cause disintegration of historic structures.

Lithographic limestone: Microcrystalline limestone made of micrite only.

Lithomonitors: Marble tombstones of the same composition or other dated objects made of carbonate rock that can preserve the reaction product of pollutants with calcite thus reveal the atmospheric composition.

Louisville limestone: Geologic formation of Silurian age widely outcropping in the Louisville, Kentucky, area. Within the Louisville limestone is a thick bed termed Big Blue, which is a fine-grained dolostone.

Magma: Molten rock from which primary silicate minerals crystallize on cooling. Magma that reaches the earth's surface is referred to as lava.

Marble: A highly compact, nearly nonporous rock made of large interlocking crystals of calcite or dolomite. Some of the types of marble discussed in this book are Alabama marble, Carrara marble, Georgia marble, Makrana marble, Penetelic marble, and Vermont marble.

Mass-transfer coefficient: See external diffusion.

Mass-flow controller: Device to deliver constant volume of air used in making an artificial atmosphere.

Mechanical behavior: Refers to mechanical properties of stone, such as compressive strength, tensile strength, and abrasive index.

Mechanical weathering: Weathering caused by stresses produced within the rock, such as frost action and salt crystallization.

Meniscus: Surface of the water table in a capillary. The meniscus is concave upwards, indicating water suction, in carbonate capillaries because of the larger attraction to water by the capillary than the inter-water molecules.

Mercury porosimetry: A method to determine pore size and pore volume in a solid by forcing mercury into the pores under pressure. While the volume of mercury intruded is measured as capacitance in picofarads and converted to cm^3/g, the radius of pores r (μm) is determined by $r = (2\gamma \cos \theta)/P$, where γ and θ are the surface tension (485 dyn/cm) and the contact angle of mercury (130°), respectively, and P (psia) is the applied pressure.

Micrite: Fine-grained carbonate mud and matrix in carbonate sediment and rock, respectively.

Micron: Unit of length equal to one-millionth (10^{-6}) of a meter, usually denoted by the symbol μm.

Mineral: Naturally occurring element or compound with a precise chemical formula and a regular internal lattice structure. Organic products are usually not included.

Modeling decay rates: Developing mathematical relation between factors that control decay (diffusion, mass transfer, kinetics, etc.) and the experimentally measured decay rate.

Modulus of elasticity (E): Modulus of elasticity, or Young's modulus, is the ratio of stress to strain in the domain of elastic deformation. The modulus of elasticity expresses material resistance to structural deformation.

Mohs' scale of hardness: Empirical scale of mineral hardness with talc as 1, gypsum 2, calcite 3, fluorite 4, apatite 5, orthoclase 6, quartz 7, topaz 8, corundum 9, and diamond 10.

Mollusca or mollusks: Phylum of invertebrate organisms, which includes the cephalopods, gastropods, and pelecypods.

Mortar: Masonry material used between the dimensional blocks or bricks. The main purpose of mortar is to drain water away that may be trapped behind the masonry construction.

Nanobacteria: Small bacteria (0.05 μm to 0.2 μm) considered responsible for deposition of carbonate sediments.

Natural environment: Environment that lacks the industrial affluents sulfur dioxide and nitrogen dioxide but contains other atmospheric gases and water vapor. (See atmosphere.)

Natural rain: See clean rain.

Neurons: Fundamental units of information storage and processing in the human neural system. (See artificial neural network.)

Nitrocalcite: See epsomite.

Noncarbonate minerals in carbonate rocks: Mostly clay minerals, quartz, and water-soluble salts.

Normal joints: Fractures in rock due to uplift or tension that dip in the direction of downthrow of strata, or are vertical to the earth's surface.

Nummulites: Large discoid foraminifers composed of certain limestones of Eocene age, such as the Mokattam formation from which the Great Sphinx of Giza is carved.

Octahedral sheet: Structural form in clay minerals.

Octahedron: Geometric form assumed to have an aluminum or magnesium atom surrounded by six oxygen atoms and hydroxide anions.

Orthochemical: In earlier concepts of carbonate origin, the orthochemical component of the sediment was considered to form as a chemical precipitate as opposed to shell fragments, which were considered as biochemical.

Oolite: Sedimentary carbonate particle in the form of spherical grains precipitated from warm ocean water on carbonate platforms. Oolitic limestone is a rock composed of such particles.

Oxidation: Chemical reaction in which electrons are lost from an atom and its charge becomes more positive. Also, a chemical reaction whereby a material changes to an oxide. For example, oxidation of iron produces iron oxide that, due to its larger volume, breaks up the rock.

Packed biomicrite: Limestone containing more than 10% allochem. The upper highly durable layers of the Sphinx limestone are made of packed biomicrite.

Paleoslope: Ancient structural slope on which new sediment is deposited. For example, the substrate of the Mokattam formation sloped southward at a low angle. This is why the Sphinx strata are somewhat inclined southwards.

Percolation model: Mathematical model to determine magnitude of reaction based on change in porosity.

Performance criteria: Specifications for conservation treatment.

Period (geologic): Most commonly used unit of geologic time, representing the subdivision of an era.

Permeability: Ability of a rock to transmit fluids through pores.

Permeation tube: Device to deliver a constant volume of pollutant gases, such as SO_2 and NO_2, whereby an artificial atmosphere can be created accurately.

Petrographic microscope: Optical microscope with polarizing filters designed for identifying minerals from thin sections by studying crystallographic properties.

Petrology: Study of the composition, structure, and origin of rocks.

Plastic deformation: Deformation that produces large strains at constant stress without fracturing.

Polluted environment: Environment that contains anthropogenic gases such as oxides of sulfur and nitrogen.

Polymorph: Chemical compound that appears in two or more mineral structures. For example, calcite and aragonite are polymorphs of calcium carbonate.

Pore model: See percolation model.

Pore potential: In mercury porosimetry, the difference in work between injecting and extruding mercury or the sum of the work in entrapment and the hysterisis loop.

Pore-size distributions: Volume of pores in different pore-size ranges. Often determined by mercury porosimetry.

Porosity: Sometimes called bulk porosity, it is the percentage of the total volume of a rock occupied by pore space, or the ratio of the total pore volume to the volume of the sample.

Power law: A distribution, such as a certain pore-size distribution, is considered to obey a power law when the x- and y-axis data plotted as logarithms describe a straight line. The slope of the line is considered as the fractal dimension.

Ppb or ppm: Abbreviation for *parts per billion or parts per million*. Often used to express low concentration in the atmosphere of gases such as SO_2 and NO_2.

Pre-industrial environment: Environment of, say, prior to 1850, when the carbon dioxide content of the atmosphere was less than 280 ppm and SO_2 and NO_2 were practically absent.

Primary silicate minerals: Minerals that crystallize by the cooling of lava or magma (Table 2.1).

Projection: Representation of three-dimensional earth features on planar surfaces; the maps. Some of the types of projections are stereographic projection and equal-area projection.

Pycnometer: Device to measure the volume of a solid accurately.

Rate constants: Coefficients of parameters used for predicting rates in model equations.

Rate curve: Plot of experimentally determined conversion against time.

Recrystallization: Growth of new mineral grains in a rock at the expense of old grains that supply the material.

Relative humidity: Ratio of the amount of water vapor in a given volume of air to that needed to saturate this volume at the same temperature, expressed as percentage.

Rhizosphere: Acid environment around plant roots that promotes decay of carbonate rocks.

Saturation coefficient: The ratio of volume of water that can saturate a rock sample by capillary force to the saturation volume of water when forced into the rock by external pressure. The value of the saturation coefficient is often less than 1, but those rocks with a saturation coefficient of 0.9 or less are considered durable.

Sedimentary rock: A rock formed by the accumulation and cementation of mineral grains transported by wind, water, or ice to the site of deposition or chemically precipitated at the depositional site.

Seismic discontinuity: Abrupt change of seismic wave velocity at the interface between two layers of different density.

Seismic profile: Data collected from a set of seismographs arranged in a straight line with an artificial seismic source, especially the times of P-wave arrivals.

Seismic refraction: Mode of seismic mapping in which the profile of the subsurface can be obtained by examining the velocity of artificially generated earthquake waves from near-horizontal strata below the surface. For example, weathered layers will have low velocity due to their lower density.

Shale: Very fine-grained detrital sedimentary rock composed of silt and clay.

Shore hardness: Hardness measured by the Shore scleroscope that indicates values of hardness by dropping a diamond ball on sample surface.

Shrinking unreacted core model: Model to evaluate the magnitude of change of the original stone if the gas penetrates into the sample. Used in this book, it refers to the migration of calcium ions to the surface.

Silt: Clastic sediment in which the mineral particles, mostly quartz, are between $\frac{1}{16}$ mm (64 μm) and $\frac{1}{256}$ mm (4 μm) in diameter.

Solubility: Mass of a substance that can be dissolved in water at chemical equilibrium.

Space lattice: Indefinite repetition of atoms in space producing rows and planes in a unit cell.

Sparse biomicrite: Limestone made largely of fine-grained particles and containing less than 10% allochem. The less durable lower layers of the Sphinx limestone are made of sparse biomicrite.

Spheroidal weathering: In a conventional geological sense, it is the formation of spherical residual inner cores by the weathering of boulders. In this book it refers to the rounded profile of a surface of an outcrop produced by weathering.

Sphinx: The Great Sphinx of Giza is a monolithic structure carved in the limestone. The term *Sphinx limestone* refers to this limestone, which is of Eocene age and belongs to the Mokattam formation.

Strain: Fractional change in length or volume when a solid is subjected to stress.

Stratification: Structure of sedimentary rocks that have recognizable beds. Stratification is parallel bedding when the bedding planes are parallel and cross-bedding when some strata are inclined with respect to the over- and underlying strata.

Stratum contours: Imaginary lines representing the same elevation on a paleoslope.

Stress: Force that may produce strain in a rock.

Strike: The angle between North and the horizontal line contained in any planar surface (inclined bed, fault plane, etc.); also the geographic direction of this horizontal line.

Structural deformation: Change in attitude from horizontality and fractures in rock layers caused by mechanical stress due to tectonic movement.

Stylolites: Zigzag dark lines in a carbonate-rock outcrop subparallel to bedding.

Supersaturation: Unstable state of a solution that contains more solute than its solubility allows.

Surface chemical reaction: Portion of the chemical reaction controlled by kinetics, such as the buffering capacity of carbonate rock vis-à-vis the pollutant; expressed in cm/h.

Surface reduction or recession: Loss of surface relief of carbonate surfaces caused by solution in rain. It can be measured through determining calcium ions in rain runoff from samples and expressed as equivalent mass or thickness of original calcium carbonate lost.

Surface tension: Large attractive force on the surface of the drop of water resulting from intermolecular forces within the drop.

Surfactant: Surface active agents. Normally the term is used for media that permit deep penetration of fluid into pores but is also used for media that can change the surface chemistry and thus alter the chemical reactivity of carbonate rock.

Swelling clays: Also called expanding clays. Clays that can adsorb water between layers, thereby enlarging the spacing d when wet.

Synaptic strength: In the human neural system, the strength of the interaction between neurons; represented by *weights* in the artificial neural networks.

Taj Mahal: Constructed from 1631 to 1653, the marble of the Taj Mahal is chemically well preserved due to absence of industrial pollutants in the area.

Tectonics: Movements and deformation of rocks, due to epeirogeny, folding, faulting, and plate tectonics.

Tensile strength: Force at which a material will rupture when subjected to tensional force.

Tensional stress: Force with which a material is pulled apart.

Tennessee marble: A compact fossiliferous limestone, but called marble in commerce due to its use as ornamental stone.

Tetrahedral sheet: Sheet structures in clay minerals in which the tetrahedra are connected to each other through oxygen atoms. The arrangement of the tetrahedral and octahedral sheets is used to classify clay minerals.

Tetrahedron: Geometric form of four equal triangular sides, modeled as containing four oxygen atoms around a silicon atom producing the complex ion or radical $(SiO_4)^{4-}$, which is the fundamental building block of primary silicate minerals.

Texture: In carbonate sediments and rocks texture refers to the characteristics of grain or crystal size, size variability, rounding or angularity, and preferred orientation.

Thermal expansion: Property of an increase in volume as a result of an increase in internal temperature; often expressed as fractional change in length per °C.

Thermodynamic work: In mercury porosimetry the PV work, often expressed as joules/gram.

Thin section: Glass slides with nearly 30 μm thick slivers of mineral or rock used for identification of minerals by light microscopy.

Travertine: Freshwater limestone formed in caves and around hot springs; called travertine marble in the stone industry.

Ultraviolet radiation: Short-wavelength radiation that disintegrates polymers that readily absorb it.

Unconsolidated material: Soft sediment that has no mineral cement or matrix binding its grains.

Unit cell: Smallest volume of a three-dimensional repetitive pattern of atoms that contains a complete sample of the atomic structural units in a mineral that compose the pattern.

Uplift: Broad and gentle epeirogenic increase in the elevation of a region without a eustatic change of sea level.

Verde antique: A dark-green rock made essentially of serpentine, a silicate mineral, but commercially called marble because it is used as an ornamental stone.

Volcano: Any opening through the crust that has allowed magma to reach the surface, including the deposits immediately surrounding this vent.

Wavelength: Distance between two successive peaks, or between troughs, of a cyclic propagating disturbance. X-ray energy can be visualized to travel in waves. The wavelength (λ) of copper $K\alpha$ radiation often used to obtain X-ray diffraction patterns is 1.5148.

Water-soluble minerals: Often called evaporites, these minerals become incorporated in the carbonate sediment and eventually become part of the limestone.

Weathering: Processes that decay and break up rock when exposed to atmosphere, by a combination of physical disintegration (mechanical weathering) or chemical decomposition (chemical weathering).

Weights: In artificial neural networks the $n \times p$ matrix is multiplied by weights to train ANN. These weights are supposed to mimic the synaptic strength in a biological neural network.

Wetting: Adsorption of water on pore walls by capillary action. Carbonate rocks are hydrophilic; they wet readily on contact with water.

X-ray diffraction: Process of identifying mineral structures by exposing crystals to X rays and studying the resulting diffraction pattern.

X-ray photoelectron spectroscopy: Called XPS, it is a highly accurate method to determine elements when evaporated by plasma from a sample surface.

◼ SUBJECT INDEX

Page numbers in *italics* designate figures; "t" designates tables. *See also* refers to related topics or more detailed topic breakdowns.

273